Collins

INTERNATIONAL
LOWER SECONDARY

Computing

Teacher's Guide

8

Laura Sach

William Collins' dream of knowledge for all began with the publication of his first book in 1819.

A self-educated mill worker, he not only enriched millions of lives, but also founded a flourishing publishing house. Today, staying true to this spirit, Collins books are packed with inspiration, innovation and practical expertise. They place you at the centre of a world of possibility and give you exactly what you need to explore it.

Collins. Freedom to teach

Published by Collins

An imprint of HarperCollins*Publishers*
The News Building, 1 London Bridge Street, London, SE1 9GF, UK

HarperCollins*Publishers*
Macken House, 39/40 Mayor Street Upper, Dublin 1, D01 C9W8, Ireland

Browse the complete Collins catalogue at collins.co.uk

10 9 8 7 6 5 4 3 2 1

ISBN 978-0-00-868409-9

British Library Cataloguing-in-Publication Data
A catalogue record for this publication is available from the British Library.

Author: Laura Sach
Publisher: Catherine Martin
Product manager: Saaleh Patel
Project manager: Just Content Ltd
Development editor: Julie Bond
Copyeditor: Paul Clowrey
Proofreader: Laura Connell and Tanya Solomons
Cover designer: Gordon McGilp
Cover illustrator: Amparo Barrera, Kneath Associates
Internal designer: Sam Vail
Illustrations by Jouve India Ltd
Typesetter: Sam Vail
Production controller: Lyndsey Rogers
Printed and bound by Ashford Colour Press Ltd

MIX
Paper | Supporting responsible forestry
FSC™ C007454
www.fsc.org

This book contains FSC™ certified paper and other controlled sources to ensure responsible forest management.

For more information visit: www.harpercollins.co.uk/green

Acknowledgements

The publishers gratefully acknowledge the permission granted to reproduce the copyright material in this book. Every effort has been made to trace copyright holders and to obtain their permission for the use of copyright material. The publishers will gladly receive any information enabling them to rectify any error or omission at the first opportunity.

We are grateful for the following for permission to reproduce their images:

p.95 iconation/Shutterstock, p.95 Turkan Rahimli/Shutterstock.

Support materials and screenshots are licensed under the Creative Commons Attribution-ShareAlike 2.0 license. We are grateful to the following for permission to reproduce screenshots. In some instances, we have been unable to trace the owners of copyright material, and we would appreciate any information that would enable us to do so.

Python Software Foundation: Permission obtained to use screenshots demonstrating Python programming language features.

Scratch Foundation: Authorised usage of screenshots showcasing Scratch programming environment elements.

EduBlocks: Granted clearance for the inclusion of screenshots depicting EduBlocks interface and functionalities.

Scratch is developed by the Lifelong Kindergarten Group at the MIT Media Lab.

PSF's License Agreement and PSF's notice of copyright, i.e., "Copyright (c) 2001 Python Software Foundation; All Rights Reserved" are retained in Python 2.0.1 alone or in any derivative version prepared by Collins (Licensee).

Contents

Teacher's Guide

Introduction

Stage 8 Computing builds on the foundation set in previous stages, building on familiar themes but expecting students to learn new concepts and skills. They will transition from using Python as a block based language in Stage 7, to writing their own Python code in text. Students will explain the journey a message takes from the point it is typed, travelling via networks to the other side of the world and linking together various concepts on the way. They will also learn about real-world applications of technology, such as a database in a library and applications of AI, and they will explore how radio technology is used in the workplace, ultimately creating a helpful device using the BBC micro:bit.

Overview

The Collins series is built around six themes relating to computing and digital literacy. These are:

- Our digital world – Providing the tools and insights into safely navigating the digital world around us
- Content creation – Creating content using a variety of software and hardware, including office tools and multimedia devices such as tablets
- Create with code – Providing the fundamentals of programming and computational thinking skills
- How computers work – Lifting the lid on the specific technologies that make up a computer system
- Connect the world – Understanding how the world is connected through networks, the internet and the World Wide Web
- The power of data – Collecting, analysing and presenting data linked to real-world activities that create or empower change

Stage 8 has eight chapters that relate directly to these six themes, with three chapters dedicated to creating with code.

Stage 8 Kit list

Chapter 1 – Our digital world	• Printed copies of 'Social media post' worksheet (optional) • Desktop publishing or word processing software • Check whether your network settings will allow you to inspect cookies (used in Lesson 2) • Spreadsheet software • AI image generator • Software that allows graphics and text to be manipulated – this might be desktop publishing software (either app based such as Publisher, or online such as Canva), but some word processing and presentation software also has suitable capabilities
Chapter 2 – Content creation	• An example of a virtual tour (for example of a local attraction or museum) • Presentation software • Access to use a search engine

	• Multimedia capture devices, such as camera, video recorder or microphone. Depending on your school's policies, students could use their own mobile phones. • (Optional) Video editing software • A device that supports augmented reality apps, such as a mobile phone or tablet • Printed copies of 'Awards' worksheet
Chapter 3 – Create with code 1	• Devices capable of running Python and a Python IDE • Scrap paper • Pencils
Chapter 4 – How computers work	• Coloured pencils or pens • Flowchart software (see recommendations) • Mini whiteboards or cards • Printed copies of 'Firewall rules' worksheet • Printed copies of 'Troubleshooting tool test scenarios' worksheet
Chapter 5 – Create with code 2	• Devices capable of running Python and a Python IDE • Mini whiteboards (optional) • A logic gate simulator program (see recommendations) • Starter code for quiz • Printed copies of 'Quiz feedback' worksheet • Scissors (optional)
Chapter 6 – Connect the world	• Markers – washable • Stickers (optional) • A4 paper • An ethernet cable (optional) • String • Beads, tube-shaped pasta or straws in two different colours • Printed copies of 'ASCII values for common characters' worksheet • Printed copies of 'Unicode' worksheet • Printed copies of 'Echo check messages' worksheet • Calculators (optional) • Large sheets of paper • Marker pens or felt tips in six different colours • Coloured pencils or pens
Chapter 7 – The power of data	• A3 paper (optional) • Sheets of paper or card with 1, 2, 3, 4, 5, 5 written on them • Data capture form software (see recommendations) • Spreadsheet software
Chapter 8 – Create with code 3	• Presentation software • One micro:bit per student (if this is not possible, the lesson can be completed using the online emulator) • Online access to MakeCode editor

Pedagogy

Our overall approach

Our approach to teaching is based on the latest research in ensuring that computing classrooms are inclusive environments that foster greater participation of underrepresented groups in computing and technology.

Each chapter is carefully organised to develop essential knowledge and skills whilst working towards the creation of a final project. The final projects are designed to boost creative skills and give students the opportunity to make decisions and develop outputs that matter to them. All chapters end with a showcase lesson that gives students the opportunity to develop their presenting skills, whilst gaining valuable feedback on their work.

Every lesson plan states the key pedagogy that is used in that lesson. Stage 8 uses the following key pedagogies:

Classroom talk

Classroom talk is the communication that happens in a classroom, between teacher and student, student and teacher, and between students. Discussion and social interaction plays an important role in computing education. Throughout every lesson in Stage 8 you will find 'Discuss' activities. You are encouraged to incorporate topics that have been generated by students and encourage them to share their ideas and perspectives and to ask any questions they have.

Benefits:

- Classroom talk provides an opportunity for multi-way knowledge sharing and building a shared understanding.
- Valuing student voices and perspectives increases their sense of connection to computing.

Jenkins, 2017. Classroom Talk and Computational Thinking DOI: (Website IntR2)

Get creative

This key pedagogy is used to encourage students to be creative with code or technology and provides them with the freedom to make their own design choices. Stage 8 provides a wealth of opportunities to get creative with technology, from contributing to a piece of collaborative pixel art to building a 'classroom welcome' device.

Benefits:

- Students create projects that are culturally relevant to them, leading to greater engagement.
- Providing more choice in the projects that students create leads to a greater sense of agency, particularly amongst girls.
- Spending more time on the creative elements of computing leads to greater engagement of underrepresented groups.

Steve Human et al, July 2022. (Website IntR3).

Sue Sentance et al, August 2021. (Website IntR4)

Real-world contexts
Lessons make regular links to everyday life scenarios where students may encounter the concepts they are learning about. Links are also made to other curriculum topics. In Stage 8, students learn about how AI is used in their everyday lives and how simulators are used in industry.

Benefits:

- Using real-world contexts aids students' understanding by connecting computing concepts to their everyday lives.
- Real-world contexts demonstrate the value of computing, which engages a more diverse group of students.
- Real-world contexts show the practical application of computing and make links to the world of work.

Unplugged
Unplugged activities demonstrate computing concepts without using computers. In Stage 8, students create a pixel art image in an unplugged activity.

Benefits:

- Unplugged activities have been demonstrated to positively affect students computational thinking skills.
- Unplugged activities can increase students' engagement with concepts and develop deeper understanding.
- Unplugged activities can be helpful where the amount of time that students have access to computers is limited.

Chen et al, 2023. Fostering computational thinking through unplugged activities: A systematic literature review and meta-analysis (Website IntR5).

Develop practical skills
To be effective in computing, students need to develop practical skills such as using a keyboard and mouse and using IT tools.

Benefits:

- Developing practical skills enables students to become effective users of technology.
- Developing practical skills means that students have more capacity to focus on the new computing concepts they are learning.

Showcase
Showcasing is presenting a project to an audience. In Stage 8, every chapter ends with a showcase of the work that the students have completed.

Benefits:

- Showcasing requires students to reflect on their work so that they can explain it to others.
- Showcasing develops important communication skills.
- Showcasing enables students in the audience to develop the ability to give feedback.
- Showcasing provides an opportunity to celebrate the achievements of students and the tangible outcomes of computing work.

Building on prior knowledge
Building on prior knowledge involves recall of existing knowledge of programming when learning a new programming language. Similarities can be beneficial to conceptual transfer, though attention must be

paid to explaining differences between programming languages. In Stage 8, students use their knowledge of Scratch and EduBlocks as they learn the Python text-based programming language.

Benefits:

- It enables students to acquire new knowledge of programming concepts more rapidly by building on existing knowledge (attention must be paid to differences to avoid confusion).
- It makes new programming languages less intimidating due to familiarity with similar concepts.

Ethel Tshukudu et al, 2020. Understanding Conceptual Transfer for Students Learning New Programming Languages. (Website IntR6).

Investigate

Investigate is the 'I' in the PRIMM (Predict, Run, Investigate, Modify and Make) methodology for learning programming. It involves students interacting with code in a variety of ways to deepen their understanding. In Stage 8, students carry out exploratory activities and answer questions to develop their understanding of new programming concepts.

Benefits:

- Investigation builds deeper understanding of coding concepts through interacting with code in different ways.
- It promotes discussion of questions that arise when interacting with code.

Sue Sentance et al, 2019. Teachers' Experiences of using PRIMM to Teach Programming in School. (Website IntR7).

Assessment

The books are designed to follow a mastery learning approach where all learners are supported and challenged to meet the age related expectations of the curriculum. Formative assessment should be used regularly to check the overall understanding of your class. Each chapter has a final project that is assessed summatively. Appropriate scaffolding is provided to allow for the majority of learners to meet expectations.

Formative assessment in the student workbook

The student workbook provides opportunities for learners to apply their learning to various activities. The outcome of these tasks can be used as part of a formative assessment process to measure the learner progress throughout the chapter. Where a workbook question has a specific answer, answers have been provided. There are also 'model answers' to guide your assessment needs.

Formative assessment and the lesson plan guidance

Each lesson plan has a guidance column. Where appropriate, additional support has been provided for the type of answers that you are looking for when questioning the learning. This will help to guide your formative assessment. If learners are struggling to understand the concept then you should revisit it with the support of the scaffolding and guidance provided.

Summative assessment and the project rubrics

Every chapter has a final project that the learners are working towards. Each project has a project brief with requirements for learners to meet. The projects are then summatively assessed using a rubric. Each rubric has three columns 'Working towards', 'Meets expected', and 'Exceeding expected'. The majority of the class should be able to achieve the 'Meets expected' criteria. Some learners will exceed this expectation and go a step further than required by the project brief. A few learners may fall into the 'Working towards' category and should receive additional support and scaffolding.

Learners should have access to the rubric at the start of a project so that it is clear how they will be assessed. This also helps to support their metacognitive skills.

Summative assessment and showcasing

Each chapter ends with a showcase which provides an opportunity for summative assessment. Teachers can complete the assessment rubric during showcase lessons. Teachers can ask additional questions of learners to establish their understanding of specific criteria.

Differentiation

Projects are differentiated by outcome. All learners should be aiming to 'meet expectations' for the final project. Some may exceed expectations and a small few may be categorised as 'working towards'.

Scaffolding

Activities have been appropriately scaffolded using a variety of techniques. Examples include step by step instructions, worked examples of code and partially created projects to help learners get started.

The guidance column in the lesson plan also provides specific, tailored examples for how to scaffold learning where necessary.

Build on this

Every learner should be aiming to 'Meet expectations' and complete the activities to a good standard. Scaffolding is provided to support this. There are some learners that will have a desire to go above and beyond and exceed expectations. Every lesson has a 'build on this' section that includes guidance on how learners can stretch their learning further.

Hardware and software guidance

EduBlocks

EduBlocks is a programming tool that allows users to learn Python by dragging and dropping blocks. EduBlocks allows students to build on the skills they learned from programming using Scratch.

EduBlocks can be accessed via (Website IntR5).

Full information about setting up accounts and classrooms can be found on the EduBlocks website: (Website IntR6). This includes instructions for teachers to set up classrooms and instructions for students to join classrooms.

EduBlocks also provide 123 Coding Cards, which map Scratch to EduBlocks to Python: (Website IntR7). These can be printed as classroom resources or accessed online.

Scratch

Scratch is a free graphical block based programming language developed for young people. Scratch can be used online through a web browser on a laptop, desktop computer, or tablet. Alternatively, Scratch can be used without internet connection by downloading and installing the offline version it on a laptop or desktop computer. It is important to decide which option is best for you and practise using that method before a lessons with students.

Scratch can be accessed via (Website IntR8).

Full information about Scratch accounts can be found on their website (Website IntR9). This also links to specific information about teacher accounts and using Scratch offline.

Machine Learning for Kids

Machine Learning for Kids is a free tool that adds machine learning capabilities to the Scratch programming environment.

Machine Learning for Kids is an online website and tool available at (Website IntR10) and can be accessed from a web browser. An internet connection is required to use the website. No installation is required to use Machine Learning for Kids however, we recommend trying out the lessons before using it within a class, in case you have any problems accessing the website.

Machine Learning for Kids has two major parts:

1. A web-based tool where you can set up a machine learning model and work with training data.
2. A version of Scratch with extension blocks for working with a machine learning model.

You can optionally create accounts for students on Machine Learning for Kids. Accounts are not required for the lessons in Chapter 7. Students can complete the whole Chapter using the 'Try it now' feature. Using accounts offers additional features, including saving machine learning projects with training data.

If you do decide to create accounts for your students, you should follow any permissions processes required by your school and country.

Note that Machine Learning for Kids uses a customised version of Scratch.

You can find more information about Machine Learning for Kids, including how training data is stored at (Website IntR11).

Note that Chapter 7 only uses numerical data. No personal information from students is used in training the machine learning models.

Online word processor with collaborative capability

We have used Google Docs as an example tool, but you may use a different tool already within your school. The key feature is that it must allow real-time collaborative working from students at different devices and enable them to both write comments and track changes.

Google Docs can be accessed via (Website IntR12).

Students must have a way to share created documents with others, either with a shared storage area or by using the URL.

Full support for using Google Docs can be found at (Website IntR13). Select 'Google Docs Editors', and then 'Google Docs'.

Spreadsheet software

The Spreadsheet software used is Google Sheets, but you can use any spreadsheet software that enables you to add conditional formatting. One advantage of using Google Sheets is that you can use Google Forms when capturing data and have it automatically populate a spreadsheet. If you choose other software this may not link as seamlessly.

Google sheets can be accessed via (Website IntR14).

Full support for using Google Sheets can be found at (Website IntR15). Select 'Google Docs Editors', and then 'Google Sheets'.

Data capture form software

Google Forms is used as the data capture form software. You can use any online forms software, but ideally it should be able to automatically feed your spreadsheet software.

Google Forms can be accessed via (Website IntR15).

Full support for using Google Forms can be found at (Website IntR16). Select 'Google Docs Editors', and then 'Google Forms'.

Micro:bit

A micro:bit is a programmable device used in physical computing.

Students will use the EduBlocks environment to create Python code for their micro:bits. They will test the code using the simulator in EduBlocks before connecting their micro:bit to the computer and downloading the code to the micro:bit. Once the code is on the micro:bit, students can use it connected to their computer or can remove it and instead power it from a portable battery pack.

There are two versions of the micro:bit. The newer version has a processor, microphone, 25 programmable LED lights, 2 input buttons, a touch logo, GPIO pins, a speaker, a radio antenna, a compass, an accelerometer and a battery socket. The newer version is used in these books, although the original version can be used without sound options or alongside small electronic components to add sound.

A getting started video and text-based guide can be found on the micro:bit website (Website IntR17). There are also short CPD courses available for teachers at (Website IntR18).

Full features are shown via (Website IntR19).

Chapter 1 – Our digital world

Project: Targeted advertisement

Chapter overview

Have you ever wondered how advertisers know so much about you? In this chapter, students will learn strategies to evaluate advertisements and other online information, and they will understand the data trail they leave as metadata. Then, they will create a data model to help them to formulate a new lunch product before advertising it to their classmates.

Chapter 1: Lesson summary		
Lesson	**Learning Objectives**	**Summary**
1	• Evaluate the validity of information found online • Evaluate the suitability of a website by looking at the URL	Students will learn about the parts of a URL and what hints they give about a website. They will learn some questions they can ask to evaluate the reliability of online content, and then use these skills to evaluate a social media advertisement.
2	• Explain why usernames should not reveal your personal identity • Explain the term, metadata, and how it can be used to describe users and their activity	Students will think about how personal data can be leaked by means of a username. They will then learn about the metadata trail they leave when they visit a website and investigate what data is being stored about them in cookies.
3	• Describe how metadata can be sold as a product	Now that they understand that metadata is collected about them when they browse the web, students will learn that this metadata can have value to companies. They will practise data mining on a small scale, using an example data set.
4	• Identify key features of models such as simulators, including their data requirements • Use a spreadsheet to model a real-life scenario, using what-if analysis	Companies do not only use data collected about users to inform their decisions. Modelling of different scenarios helps companies to decide on which products to launch. Students will create a data model to help them to devise a new lunch product.
5	• Use sound, video, text and other multimedia to create a digital artefact	Using the skills learned about the reliability of information in the first part of this chapter, students will create a social media advertisement to advertise their product idea to their peers.

6	• Use a spreadsheet to model a real-life scenario, using 'what if' analysis	The product advertisements are showcased to the class. Each class member is invited to select multiple products to purchase. Students then use the class data and their models to answer various 'what if' scenarios.

End of chapter project	Targeted advertisement
Example ideas that learners could design:	• Social media advert for a new lunch product aimed at peers • Data model for ingredient costings, capable of modelling different 'what if' scenarios

Why we are learning this	
Useful skills covered in this chapter	• Ability to evaluate information found online • Using spreadsheet functions to create a model
Links to future careers	• Business terminology – cost, revenue, profit • Using a spreadsheet for business planning • Social media and marketing
Links to other subjects	• Food and Nutrition – cost of food ingredients • Business Studies – advertising, costs, revenue, profit • Mathematics – percentage increase and decrease

Chapter 1 – Our digital world

Project: Targeted advertisement

Name	

	Working towards (1)	Meets expected (2)	Exceeding expected (3)	Score
Data model	• Collects suitable data to assist with the choice of a new lunch product and displays this in a spreadsheet to form a basic data model	• Creates a suitable data model to help to devise their chosen new lunch product	• Creates a more complex data model that contains information relevant to their chosen lunch product	
Use of text, images and sound	• Creates a simple social media advertisement to advertise their lunch product idea to their peers that uses text and one AI generated image	• Uses text, AI generated images and sound to create a visually appealing advertisement for their lunch product	• Is creative with their use of AI generated images, text, sound and videos to create an engaging social media advertisement	
What-if scenario	• Uses a spreadsheet to model what happens when a change is made to one of the items in their data model	• Uses their spreadsheet model to investigate the effect of the new scenario they are presented with	• Uses what-if analysis effectively to compare alternative scenarios	

Teacher feedback	
Learner response	

Chapter 1:	Lesson 1 Evaluating online information
Student's Book page 2	
Learning Objective/s	• Evaluate the validity of information found online • Evaluate the suitability of a website by looking at the URL
New keywords	**Uniform Resource Locator (URL)** – a unique address that specifies the location of a website
You will need	• Printed copies of 'Social media post' worksheet (optional)

Activity	Instruction	Guidance
1	• Discuss 1, (Page 2) Ask students to look at the image of the advert and discuss with a partner – would you enter the competition? • Bring the class back together and ask a variety of pairs to give their opinions and reasons why or why not.	• Students may be quick to assert that they would or would not enter the competition, but make sure to prompt them to think about the reasons why or why not. What is it about the advert that caused them to come to the conclusion? You could prompt students by asking whether there could be any reason to suspect this competition might not be genuine. Some answers may include: • Not a known website • Spelling mistakes • Poor grammar • Timer – prompting you to act quickly

Activity	Instruction	Guidance

2	• Talk through the ways you can evaluate the reliability of a website by means of the URL (📖 Pages 2 and 3).
	• Ask students to complete (📖✏ Page 1) Task A, 'Evaluate the URL'. Ask students to write down any information they can deduce about the three URLs in the Workbook.
	• Bring the class back together and ask for answers and feedback on the Workbook task. Discuss any points where students did not agree, for example:
	    • Is the Canada website really the official website of Canada?
		• The URL includes the protocol (HTTP or HTTPS for a webpage). Students will study protocols in more detail in Stage 9. They are not expected to understand this beyond the fact that HTTPS is a secure transmission at this stage.
		• If students are stuck on the activity in the Workbook, prompt them to look at the different areas highlighted (📖 Pages and 3):
		    • Does it use HTTP or HTTPS?
		    • What is the domain name? Does it have spelling mistakes?
		    • What is the TLD and what information does that give you about the site?
		• Some countries will only allow residents or nationals of that country to purchase domains with their extension. For example, the .fr domain belongs to France, and only residents of the EU or EFTA can purchase a .fr domain name.

Activity	Instruction	Guidance
3	Ask students to look at (Page 2) Task B, 'Mind map' in pairs. What answer would you be looking for if the information was reliable?	• **Purpose** – This could be journalism, entertainment, publicity or simply the thoughts of an individual. • **Search results** – Search engines have their own methods of assessing the popularity of a site. Popularity does not necessarily equal reliability, but if the site does not appear in search results that is potentially an issue. • **Have you heard of it?** – You may already know information about the source, e.g. if it is a national news broadcaster. Do they they have a reputation for reliability? • **Is it verified?** – Some social media platforms will verify high-profile users such as celebrities and businesses. However, on other platforms, verification can be purchased, so this is not necessarily an indication of reliability. • **How well written is it?** – Sources that are well written and free from mistakes are likely to have been reviewed and edited, e.g. books, pieces by journalists. This may indicate a higher level of reliability compared to, for example, a social media post by an individual.
4	• Use Investigate 1 (Page 3). Provide the social media post worksheet either as a printed copy or electronically. • Students should create a poster to teach other students how to evaluate the reliability of information found in social media posts. • To complete the task, students could use software, or print out a screenshot and annotate it by hand.	• You can either provide the worksheet as a printed copy, or as an image for students to insert into a desktop publishing program and annotate. • The purpose of the exercise here is for students to apply what they have learned about evaluating reliability to the social media advertisement provided. You will be able to assess how well they have understood the material by the quality of the explanations and justifications they give in their annotations on the poster.

Activity	Instruction	Guidance
5	• Reflection: In pairs or small groups, ask students to discuss 'What did you learn that was new?' and then complete (Page 3) 'Reflection'.	
	Homework (Optional) (Page 3) • Choose an online source of information (e.g. webpage, advertisement, email or social media post) and evaluate its reliability.	• This is an opportunity for students to practise the skills learned in the lesson within a real-world context. They could produce a piece of annotated work similar to the final task in this lesson.

Build on this: You could go into more depth about how the purpose of the information affects its reliability. For example, information that is intended to promote products from a particular brand will not mention products that might be equally as good but cheaper, sold by a competitor. Can students find information from a source with an obvious bias?

Chapter 1: Lesson 2 Usernames and metadata

Student's Book, page 4

Learning Objective/s	• Explain why usernames should not reveal your personal identity • Explain the term 'metadata' and how it can be used to describe users and their activity		
New keywords	**Cookie** – a small file stored on your computer by a website **Metadata** – 'data about data', for example who created the data, when it was created, and what device it was created on		
You will need	• Check in advance that students have access to the internet and a web browser to complete Investigate 2 (Page 6).	**Key pedagogy**	Real-world contexts

Activity	Instruction	Guidance
1	• Display the examples of usernames or ask students to look at them (Page 4). • Use Discuss 2, 'What information might a username leak about the user's identity?'	• You may wish to copy the usernames from the Student's Book onto the board and/or add some more examples of your own.

Activity	Instruction	Guidance
2	• Ask students to complete (Page 4) Task A, 'Why choose an anonymous username?' Why should your username not reveal your personal identity? • Use Discuss 3: When students have completed the workbook task individually, ask each student to join up with three other students and compare lists (Page 4). • Ask groups to share their answers with the class.	• When students join up with others to compare lists, each student should read their list in turn and other students can either agree with the point if they wrote that too, challenge it if they think their peer has made a mistake, or note it down if it is not a point they thought of. All members of the group should be actively doing something during the activity. • Some example answers are provided in the answer section.
3	• Explain that metadata is collected about you when you visit a website (Page 5). • Ask students to complete (Page 4) Task B, 'What metadata is recorded?'. Students write down what metadata they think is recorded when they perform certain actions online.	• In this context, the metadata is data about your website visit that is not used for the purpose of the visit itself. For example, the device you are using might be recorded but is not used as part of your search for a dress on an online shopping website. It is sometimes difficult to distinguish between data and metadata, because as soon as metadata is used for another purpose (for example, as students will see in the next lesson, as information to inform advertising), the metadata that was collected incidentally itself becomes data. • For the 'items searched for', this means when using the search functionality provided on the website, and *not* which search terms were typed into a search engine to get to the website in the first place (although this information is also sometimes available).

Activity	Instruction	Guidance
4	• Use Investigate 2. Ask students to investigate whether they can find any metadata being stored about them inside cookies on their computer. • Students can do this by inspecting the cookies inside their browser. • Ask students to complete (Page 5) Task C, 'Data from cookies'. Students write down any interesting data that they find stored in cookies.	• Check that it is possible to inspect cookies on the browser you have available, while logged in on a student account, prior to the lesson. If it is not, you may need to run this activity as a whole-class activity and project your own account onto the board instead. • It is possible to inspect the cookies being stored on a variety of browsers. Here are instructions for some common browsers: • **Google Chrome**: right click/control click on any webpage > select 'Inspect' > click on the 'Application' tab > select 'Cookies'. • **Microsoft Edge**: Three dots menu > Settings > 'Cookies and site permissions' > 'Manage and delete cookies and site data' > 'See all cookies and site data'.
5	• Reflection: In pairs or small groups, ask students to discuss 'Are you concerned about anything online?' and then complete (Page 5) 'Reflection'.	• Students may feel concerned that they are being tracked online. Provide the opportunity for students to raise concerns with you, either via their reflections in the Workbook or in person.
	Homework (Optional) (Page 5) • Look at a metadata profile about a website user. What information about this person can you deduce from the metadata?	• If you are struggling with access to do the practical activity (Page 6) because of network restrictions, you could do this activity in class instead, and ask students to attempt the practical activity as homework if this is feasible with the access to technology available.

Build on this: Research ways you can protect yourself from leaking metadata when you browse the web. Examples could include browsing in incognito mode, choosing a username that does not leak personal information, and setting your browser software to decline non-essential cookies. There are also web browsers that limit the amount of tracking data you share, such as DuckDuckGo.

Chapter 1:	Lesson 3 Data mining		
Student's Book page 7			
Learning Objective/s	• Describe how metadata can be sold as a product		
New keywords	**Data mining** – analysing large amounts of data to find patterns and trends **IP address** – an address allocated to you by your internet service provider, which reveals your rough geographical location. An IP address looks like a set of four numbers separated with a dot, for example 172.217.22.14		
You will need	• Spreadsheet software	**Key pedagogy**	Classroom talk

Activity	Instruction	Guidance
1	• Talk through the technical reasons that metadata was originally collected (Pages 7 and 8). • Make sure to discuss public IP addresses being logged as metadata.	• Students may be concerned to learn that their geographical location is leaked via their IP address. The location is not precise; it cannot reveal your location more accurately than a city or area level. However, it does mean that although students may feel anonymous when browsing the web, it is relatively easy to trace their actions back to them. • When you search for your public IP address, you may be shown addresses in both IPv4 and IPv6 format. This is because networks across the world are transitioning from IPv4 to IPv6, which is a newer version of the protocol. You may be most familiar with IPv4 addresses, which consist of four numbers between 0 and 255, separated by dots, e.g. 127.0.0.1 • Be aware that using a Virtual Private Network (VPN) can mask your IP address from being logged and provide a fake IP from another country, but this is illegal in some countries.

Activity	Instruction	Guidance
2	• Introduce the definition of 'data mining' and how companies use metadata (Page 8). • Use Discuss 4 (Page 8). • You could either run this activity by asking students to discuss in pairs/groups and then report back to the class, or as a whole-class question and answer session. • For (Page 6) Task A, 'Why is metadata useful?', ask students to note down their findings from the discussion exercise in their Workbooks.	• There will be other answers that are correct but not listed.
3	• Go through the 'data mining' section with students (Pages 9 and 10). • Use Investigate 3 (Page 10). Provide students with a copy of the data. • Ask students to complete (Page 6) Task B, 'Interrogate the data'. Working individually, students answer the questions provided in their Workbooks.	• Students will need to use a combination of sorting and spreadsheet formulae to answer the questions about the data provided. • If you think it is necessary recap some basic spreadsheet skills with students, for example selecting a range, typing a formula and formatting cells. Instructions for how to sort data using Google Sheets are covered in the (Page 9). • Here is a hint for the type of operation the students will need for each of the questions: 1. Sort 2. Sort, then select specific cells. You can see how many cells are currently selected in the footer of the sheet. 3. Sort 4. Sort using two columns 5. AVERAGE 6. Sort 7. Sort using two columns 8. Sort, select data, then insert chart • Students might find a shortcut helpful to highlight the whole sheet.

Activity	Instruction	Guidance
4	• Reflection: In pairs or small groups, ask students to discuss how confident they are that they have understood the concept of metadata and what it can be used for. Then ask them to complete (Page 7) 'Reflection'.	
	Homework (Optional) (Page 7) • Why is it difficult to determine which item was the most searched for, using the data provided?	• This task encourages students to think about the quality of the data provided, and any limitations there might be to drawing useful conclusions from it. • Students may have previously used multiple-choice questions to categorise responses in a questionnaire. If a free-form text response is used, they are likely to get a huge variety of similar answers, which is hard to analyse. Unfortunately, search terms are always free-form, so this raw data needs to be labelled to categorise it. A more modern approach might be to designate specific keywords that might be found in search terms, and to count how many times each of these appear, as a measure of popularity.

Build on this: Ask students to use the spreadsheet software to draw additional graphs of interesting things they have found out about the data.

Chapter 1: Lesson 4 Models

Student's Book, page 11

Learning Objective/s	• Identify key features of models such as simulators, including their data requirements.		
New keywords	**Model** – a simulation of a real-life system on a computer		
You will need	• Spreadsheet software	**Key pedagogy**	Real-life scenarios

Activity	Instruction	Guidance

1	• Talk through the definition of 'model' with students (Pages 11 and 12).
	• Explain some different examples of models:
	  • energy company
	  • taxi company
	  • engineer
	• Use Discuss 5 (Page 12). Divide students into groups and allocate each group one of the models. What are the data requirements?
		• Companies do not just use data collected about users to inform their decisions. Sometimes the data model is the product itself!
		• Some of the data that students will come up with will be able to be calculated from other data. For example, the number of taxis within a given radius can be calculated if you know the location of all the taxis. As a stretch question, can students distinguish between data that can be calculated from other data, and data that is fundamental to the model?
		• There are some example answers for the discussion task in the answers document.

2	• Introduce the project brief and explain the definitions of 'cost', 'profit' and 'revenue' (Page 13). • Ask students to carry out (Page 13) Build 1. Provide students with the starter spreadsheet. Students follow the checklist in (Page 8) Task A, 'New lunch product' to add the data and the calculations to their model. • Students should test their model as they work by typing example data into it, and checking that the formulae update as expected.	• You can sometimes get confusing results with currency values because the spreadsheet program will round values to two decimal places for display, but will **not** round the actual value in the cell. • For example, if a product costs 1.50 and contains 7 servings, the result of the portion cost calculation will be displayed as 0.21 but the value used for any further calculations will be 1.50/7 unrounded. • Using Google Sheets, if you apply conditional formatting based on the value of another cell (e.g. B7), you must specify the cell as =B7 and not just B7 otherwise the formatting will not apply correctly.
3	• Reflection: In pairs or small groups, ask students to reflect on how well creating the model went, and whether they could remember and apply the spreadsheet skills they had learned in a previous stage. Then ask them to complete (Page 9) 'Reflection'.	• Students studied spreadsheets in Stage 7 and should have covered all of the skills they need to create this model. Be prepared to carry out recap demonstrations as required.
	Homework (Optional) (Page 9) • Are there any variables missing from your model?	• Some students will notice that the costs of ingredients are by far not the only costs that would be incurred by the business. There are some example answers provided in the answer document.

Build on this: Copy the tables in the spreadsheet and fill in the correct data to model another possible product idea.

Chapter 1:	Lesson 5 Create an advert		
Student's Book, page 14			
Learning Objective/s	• Use text and other multimedia to create a digital artefact		
New keywords	**Advertisement** – a notice telling people about a product **AI image generator** – a piece of software that uses artificial intelligence to generate images based on a text description typed in by the user		
You will need	• An AI image generator (links provided below) • Software that allows graphics and text to be manipulated	**Key pedagogy**	Get creative

Activity	Instruction	Guidance
1	• In this lesson, students will create a mock-up of a social media advertisement for their new lunch product. • Use Discuss 6 (Page 14). • Link back to the advert from Lesson 1 (Page 2) and what makes online information reliable.	• In the final lesson of the chapter, students will use the advert they create to encourage their classmates to select and 'buy' their product. • Some of the features of trustworthy information discussed in the first lesson will not be possible for students to achieve; for example, their 'product' being found by means of an independent search. However, they should consider their spelling and grammar, and they can include other aspects in their mock-up such as a verified account.
2	• Ask students to complete (Page 14) Task A, 'Draw a design for your advert'.	• Students may not have access to any social media networks within school, and most social media platforms require users to be 13 or older to be able to sign up for an account. An account is not required to be able to complete this task but, if you prefer, you could gather suitable screenshots of social network advertising posts in advance to provide as reference material.

Activity	Instruction	Guidance
3	• Talk through the 'create your advertisement' section and the definition of 'AI image generator' (📖 Page 15). • Introduce Build 2 where students first create an AI image of their product and then build the advertisement they have designed using appropriate software (📖 Page 15). • Ask students to complete (📖 Page 15) Build 3. Students build the advertisement they have designed using appropriate software.	• Students are asked to generate an image using an AI image generator. The availability of these tools varies by country and free usage terms also change. At the time of printing, the following AI image generators were available with many more in production: • (Website C1L5R1) (free) • (Website C1L5R2) (free, requires account) • If, for any reason, it is not possible for students to use an AI image generator, they could draw their product image instead. • The most appropriate software to create the finished advertisement would be desktop publishing software. However, depending on what is available, you may find that it is also possible to use a word processor or even presentation software to complete the task. Another option is to use an online tool, such as: • (Website C1L5R3) (free but requires account) • (Website C1L5R4) (free) • During the next lesson, students will need to look at all of the adverts that have been created. If you know that it will not be possible for the students to display these on their screens, ask them to submit their work to you and print off the advertisements before the following lesson.
4	• Reflection: Ask students to reflect on what features of their advertisement they think will be most effective, and what they would change if they had more time. Then ask them to complete (📖✏ Page 11) 'Reflection'.	

Activity	Instruction	Guidance
	Homework (Optional) (Page 11) • Students consider the benefits and problems caused by AI image generators, such as the one they used during the lesson.	• As the use of AI is becoming increasingly common, it is important for students to have a good understanding that there are also drawbacks, especially related to copyright, to its use. • As has been demonstrated during the lesson, the use of an AI generator very quickly generates a high-quality image without much effort, and does not require any artistic talent. However, some talent is required in terms of crafting a 'prompt' (the words used to describe the image that is wanted). It may be the case that skills of the future will focus on being able to accurately describe and communicate to the computer the vision for an image, rather than creating the image manually.

Build on this: Students could generate a variety of AI images and seek feedback from their peers on which is the most appealing. Real-world examples of commercial AI generated content could also be displayed and discussed.

Chapter 1: Lesson 6 'What if' analysis

Student's Book, page 16

Learning Objective/s	• Know how to use spreadsheets that are models of real-life systems, using 'what if' analysis to compare alternative scenarios.		
New keywords	**'What if' scenario** – a simulation involving a set of criteria being applied to a model, so that the result can be observed		
You will need	• Spreadsheet software • Scrap paper • Printed scenario cards (optional) • Printed advertisements (optional)	**Key pedagogy**	Get creative / Showcase

Activity	Instruction	Guidance

1	• Explain to the class how the Role play activity (Page 16) works:
	  • Each student displays their advertisement from the previous lesson on their screen.
	  • Give each student five tokens.
	  • Each token represents the student 'buying' a particular product this week.
	  • Students circulate and look at the various advertisements. They then need to choose which products to 'spend' their money on by putting a token next to the product they wish to buy.
	  • Collate the results – count the number of tokens 'spent' on each product.
	  • Students input the number of 'sales' into their model and see how much profit they made. Make a note of this amount for each student. You may wish to celebrate the 'winner' of round 1, i.e. the student who made the most profit.
		• If it is not possible to display adverts on the students' screens, print them off in advance.
		• **Ensure that each student's advertisement includes the price of the product** – this is very important for the decision-making process.
		• If your class is particularly small or large, you may want to increase or decrease the number of tokens you give to each student. A good number of tokens is roughly a quarter to a fifth of the number of students in the class, to force students to make a decision.
		• Remind students to look at all of the adverts and to choose which products to buy based on their realistic impressions of the product, **and** its price. If this were real money, what would you actually buy? Students may not buy their own products and should be warned against only buying the products devised by their friends.

2	• Explain that you can use a model to ask 'what if' questions. • Talk through some examples of 'what if' questions for the real-life models students have already encountered (Pages 17 and 18). • Ask students to complete (Page 12) Task A, 'What if scenarios'. Model the scenarios.	• For the first scenario you will need to provide students with an item of equipment that has broken (e.g. a food blender), and how much it costs to replace. We have intentionally not provided this information because your students will be working in your local currency. • Students should save a separate copy of their model before they start editing it for each 'what if' scenario. They can either do this by duplicating the tab containing the model or saving a duplicate of the document.

Activity	Instruction	Guidance

3	• Introduce Build 5 and ask students to devise and write down a new scenario on a piece of paper (Page 19).
	• Shuffle the scenarios and distribute a scenario to each student.
	• Each student then applies the scenario they have been given to their own model in the same way as they did with the scenarios in (Page 12) Task A, 'What if scenarios'.
		• Remind students to be realistic with their scenarios – if they present their classmates with extreme costs, then the second round of the game will not work as well! You may want to have some back-up scenarios on hand in case some of the ones submitted by students are unrealistic. For example:
		  • There is a discount of 20% off all meat this week.
		  • There is bad weather forecast so you will have 5% fewer customers this week.
		  • Your product has a bad review online, and you have to give away <number> of free products in a competition to build back your reputation.
		  • The government increases tax by <percentage> on all products.
		  • A competitor is doing well, so <number> of your customers buy their product instead this week.
		  • You manage to lure customers away from a competitor: add <number> of people to your number of buyers this week.

4	• Use (Page 19) Build 6. Once the new scenario has been applied, each student is allowed to make a maximum of two changes to their product offering. They should remember to reflect these changes in their advert, if necessary.
	• Use Role play 2 (Page 19). Explain that in the role-play activity students will run the same scenario as in the previous role play but with the revised advertisements. Do students still make the same choices?
		• There will not be a huge amount of time for making changes (i.e. refactoring the whole product offering), so students should stick to a maximum of two changes. These could include things such as:
		  • changing the price of the product
		  • finding an alternative ingredient, or sourcing an existing ingredient more cheaply (with evidence!)
		  • altering their advertising in some way
		  • removing an ingredient altogether.

Activity	Instruction	Guidance
5	• Reflection: Ask students to reflect on how their lunch product idea fared in the simulation, and whether they found using the model helpful and then complete (📖✏ Page 13) 'Reflection'. • Ask students to complete in the assessment checklist on (📖✏ Page 14).	• The intention of this lesson is to give students first-hand experience of using a model to answer a 'what if' question as if they were running a business, and to use that information to react to the change in circumstances. • The actions the students take will not necessarily mitigate the change in circumstances that occurs, and it might be interesting to reflect as a class on how they would feel about this if they were really operating a business. The reflection task could result in, or start off as, a whole class discussion.
	Homework (Optional) (📖✏ Pages 13 and 14) • Students are presented with some data from a bakery and asked questions to consolidate what they have learned during the lesson.	• This task encourages students to apply what they have learned to a new scenario. They will need to interpret the scenario and use their spreadsheet skills to answer the questions.

Build on this: Some spreadsheet software offers extra features to allow you to answer 'what if' questions. Students could investigate the 'goal seek' function if it is available in the software you are using.

Chapter overview

Students will use a range of IT skills to create a short virtual tour – this might be of the school, a club they attend or something interesting in your local area. Each tour will use a range of sound, images and video, and students will present it appropriately for the audience they have chosen.

Chapter 2: Lesson summary		
Lesson	**Learning Objectives**	**Summary**
1	• Describe when it is appropriate to use emojis, GIFs and memes in communication	Introduce the concept of a virtual tour. Look at different types of communication, in particular emojis, GIFs and memes, and the contexts in which it is appropriate to use them.
2	• Create document templates and explain the benefits of using them	Students will learn how to create a document template in presentation software and understand the benefits of using a template. They will tailor their design to a specific audience.
3	• Practise typing skills • Use advanced search techniques to refine search results	Students will improve their research by learning some advanced techniques to use in search terms.
4	• Identify ways to responsibly use technology safely, and how to protect wellbeing	Students will use a multimedia device to collect some material to use in their tour. As a class, they will discuss and understand the potential risks of this technology.
5	• Use sound, video, text and other multimedia to create a digital artefact	Students will use presentation software to create a virtual tour, which a user can self-navigate with a user-friendly interface.
6	• List uses for augmented reality in familiar environments, including education and entertainment	Students will showcase their tour. They will explore augmented reality and think about how it might be used in the context of a virtual tour.

End of chapter project	**Virtual tour**
Example ideas that learners could design:	• Tour of your school • Tour of an interesting place in your local area e.g. a museum

Why we are learning this	
Useful skills covered in this chapter	• Creation of document templates – can be applied to school work in other subjects • Searching effectively
Links to future careers	• Presentation software is used in a wide variety of jobs • Uses of augmented reality in medicine
Links to other subjects	• English – thinking about the audience of a piece of work • Science – medical uses of augmented reality

Chapter 2 – Content creation

Project: Create a virtual tour

Name	

	Working towards (1)	Meets expected (2)	Exceeding expected (3)	Score
Collection of suitable images and information	• Uses a multimedia device to collect some suitable material to use in their tour	• Demonstrates use of advanced search techniques to find suitable images and information to include in their tour	• Collects a range of different types of images that are engaging and relevant to their tour	
Design	• Uses presentation software to design a simple virtual tour using images collected	• Uses presentation software effectively to put together suitable information and images to represent their chosen place	• Well-designed presentation showing good choice of images that represent their chosen place	
Virtual tour	• Creates a basic virtual tour of an interesting place	• Uses sound, video, text and other multimedia effectively to create a virtual tour that a user can self-navigate	• Uses presentation software to create a virtual tour that makes a user feel that they are really there	

Teacher feedback	
Learner response	

Chapter 2:	Lesson 1 Know your audience		
Student's Book page 22			
Learning Objective/s	• Describe when it is appropriate to use emojis, GIFs and memes in communication		
New keywords	**Audience** – the people who will look at the work you produce **Emoji** – a small icon used in text to express an emotion or represent an item **GIF** – an image format that allows animation **Meme** – a familiar image, often accompanied with text, which is shared and reused on the internet **Virtual tour** – a simulation of being at a particular location, involving multimedia such as images, text and perhaps also sound and video		
You will need	• An example of a real virtual tour that students can access or can be demonstrated	**Key pedagogy**	Real-world contexts

Activity	Instruction	Guidance

1	• Introduce the idea of creating a virtual tour as the overall project theme.
	• Use Discuss 1 (Page 22)
		• Students may know what the concept of a virtual tour is but may know it by a different name.
		• Examples of virtual tours include:
		  • tour of a house for sale
		  • street view on a map
		  • tour of a museum or historical landmark
		  • tour of a school or university for prospective students
		  • tour of somewhere you cannot actually go, e.g. inside the human body.

2	• Use Investigate 1 (Page 22) Investigate 1 Ask students to find an example of a virtual tour. • Ask students to complete (Page 14) Task A, 'Features of a virtual tour'. This asks students to note down how each of the features listed is used in a virtual tour that they have found. Can they think of any other uses?	• You might want to have on hand a specific example of a virtual tour that you know your students can access. • If you know that you will not have internet access during the lesson, you could download a virtual tour in advance, or create a slideshow, and demonstrate it before students complete the activity.

Activity	Instruction	Guidance
3	• Introduce the idea of appropriate communication and knowing your audience. • Show examples of emojis, GIFs, memes (Pages 23 and 24). Discuss with the class when each would each be appropriate and when it would be inappropriate. • Ask students to complete (Page 15) Task B, 'When is it appropriate?'. Students are asked to fill in a table to describe when it is appropriate to use emojis, GIFs and memes.	• Some of the examples in the table might be context and/or culture dependent. For example, whether you should use an emoji or a meme in an article for the school newspaper might depend on what the article is about. It may not be inappropriate to use an emoji in an instant message to a grandparent, but it may be inadvisable because the grandparent may not know the significance of the emoji.

4	• In pairs, students should plan their topics, audience and an outline of the content for a virtual tour, using (Page 16) Task C, 'Plan your virtual tour'.
	• For each item of content, specify what multimedia features it will include (images, text, sound, video etc.) and think about what type of communication will be appropriate.
		• Students will need to know in advance what technology will be available in later lessons so that they can plan their work.
		   • Will they have access to video recording equipment, microphones or cameras?
		   • Will you allow students to use personal devices or only school devices?

5	• Reflection: Ask students to reflect on the way they communicate, and they will reflect on how communication has changed since they were born. Then ask them to complete (Page 17) 'Reflection'.	• This task is intended for students to complete alone, to give them time to think about the way they communicate. The different methods of communication available to students may differ. Some students might communicate with teachers via email, while others might rely on talking face to face if this is not an option the school supports. Some students will have access to mobile phones and others will not. Answers may vary, even within the class.
	Homework (Optional) (Page 17) • Find a real example of communication (e.g. an advert, a menu, a newspaper article, a social media post) and analyse who the audience is and whether the communication is appropriate for that audience.	

Activity	Instruction	Guidance
Build on this: You could find an example of communication where there is some ambiguity as to whether it is appropriate or not. For example, some large companies have an online persona where they may post GIFs or memes as part of their social media – is this appropriate? Would it be appropriate for a different company?		

Chapter 2: Lesson 2 Document templates

Student's Book page 25			
Learning Objective/s	• Create document templates and explain the benefits of using them		
New keywords	**Template** – a pre-designed format for a document		
You will need	• Presentation software	**Key pedagogy**	Design thinking

Activity	Instruction	Guidance
1	• Use Discuss 2 (Page 25). Discuss 2	• Students may have used templates in other contexts, e.g. for drawing shapes in a flowchart, for cutting out material in a sewing pattern. These uses generally have consistency in common as their main aim.
2	• Use Discuss 3 (Page 26). Which of the purposes would the leaflet template be appropriate for?	• Prompt students to consider the features of the template that have been previously listed in the Student's Book, for example: • Is the colour scheme appropriate for the context given? It may be appropriate to use bright colours to showcase a nursery, but this is probably not appropriate in tone for the charity helping starving children. • Is the font appropriately professional? Some fonts are appropriate for business documents whereas others might feel more informal

Activity	Instruction	Guidance
3	• Demonstrate how to access the template builder (📖 Pages 26 and 27) in your specific version of presentation software. Demonstrate how to change font colours and sizes, change the background and add an image. • Ask students to complete (📖✏ Page 18) Task A, 'What are the benefits of using a document template?'. Students are asked to match the features to their benefits.	• In some popular software, the theme builder may be called 'Slide master'. The 'master' terminology is outdated because of its connections to slavery and its use is being phased out in modern digital products.
4	• Ask students to complete (📖✏ Page 19) Task B, 'Draw your template design'. Each student should design and annotate their own template idea. Remind students to take into account what is appropriate for the audience they chose in the previous lesson.	• Although students will work in pairs on the finished product, they will each create a template idea and then choose one of them to continue with, or combine elements of both in the next task. • It is important that students do not design every slide of the presentation during this task. Highlight that the exemplar shown in the Student's Book uses placeholder text (e.g. "Room name") rather than the actual title of the slide.
5	• Use Build 1 (📖 Page 28). Working together with their partners, students create a template design in presentation software, using the skills they saw demonstrated earlier in the lesson.	
6	• Reflection: Ask students to consider the features of their finished template and why they make it suitable for the audience they chose. Then ask them to complete (📖✏ Page 19) 'Reflection'.	• Students should work alone on their reflections and then compare notes with their partner to see whether they had the same thoughts.
	Homework (Optional) (📖✏ Page 20). • This task asks students to find an example of a real document and then draw a design for a template that could be used to improve it. If possible, students could bring their documents to the following lesson, or stick them in their workbooks if appropriate.	• Be prepared to provide examples of documents for any students you think may not have access at home.

Activity	Instruction	Guidance
Build on this: You could show an example of a real presentation designed without a template and ask students to point out or annotate how it could be improved by using a template.		

Chapter 2: Lesson 3 Searching for information

Student's Book page 29			
Learning Objective/s	• Practise typing skills • Use advanced search techniques to refine search results		
New keywords	**Search term** – the text you write into a search engine to find what you need		
You will need	• Internet access	**Key pedagogy**	Develop practical skills

Activity	Instruction	Guidance
1	**What information do I need?** • Ask students to think about what other information they might need to gather to add to their virtual tours and where they would find this information (📖 Page 29). • Ask students to complete (📓 Page 21) Task A, 'What extra information do you need?' in pairs, where they write down what extra information they need and where this information might come from.	• Students may be already very familiar with the place they are creating a tour about; however, there may be other information that can be researched. For example, if they are creating a virtual tour of the school, they may wish to research when it was founded and who the founders were, or perhaps whether there are any famous alumni. For a landmark, they could research its dimensions, who it was created by if it is man-made, or the materials it is made out of.
2	• Demonstrate advanced search techniques (📖 Pages 29 and 30). You may want to use other examples of search terms that are directly relevant to your students. • (📓 Page 21) Task B, 'Web quest'. Students are asked to practise these search techniques by searching for the information in the exercise.	• Make sure to research the answers to the Workbook, Task B activity before the lesson as three of the answers will vary depending on your local area. The questions are deliberately designed so that the results may not be what the students intended if they are not specific enough with their query. For example, if they search for a golf club near a wood, they may get results relating to the type of golf club called a wood, rather than a golf course next to a forest! • If you wish, you could extend this activity with some more things to search for or use it an alternative homework activity.

Activity	Instruction	Guidance
3	• Use Build 2 (📖 Page 30). Ask each pair of students to divide up the list of extra information from (📖✏ Page 21) Task A, 'What extra information do you need?' then gather the extra information required. • Some information may be typed in from the students' own knowledge and some may be found via searching on the internet.	• Typing in information from the students' own knowledge is an opportunity to practise their typing skills. • Students should record the information they find in a rough document, such as a blank presentation or text editor document – they should not add it to their virtual tour yet.
4	• Reflection: Ask students if they ran into any problems when they were searching for information on the internet • How effective were the skills they learned this lesson at improving your searches? • Then ask them to complete (📖✏ Page 22) 'Reflection'.	• Although students will complete the reflection task alone, it may be worth asking if anyone would like to share their answers with the class for this task. Other students may be able to offer advice for improving search results based on their own experiences during the lesson.
	Homework (Optional) (📖✏ Page 22) • Ask students to research an additional piece of information for their virtual tour – this could be in any format, e.g. text, a digital image.	• Space is provided in the Workbook for students to make notes, but this may not be needed if their research is in digital format.
Build on this: Ask students to create a searching 'cheat sheet' poster to give advice to other people on how to search more effectively.		

Chapter 2: Lesson 4 Using multimedia

Student's Book page 31			
Learning Objective/s	• Identify ways to responsibly use technology safely, and how to protect wellbeing		
New keywords	**Multimedia** – using more than one method of presenting information, e.g. text, images, sound and/or video		
You will need	• Multimedia capture devices, e.g. camera, video recorder, microphone	**Key pedagogy**	Develop practical skills

Activity	Instruction	Guidance
1	**What is multimedia?** • Introduce the definition of 'multimedia' and talk through the examples (Page 31). • Ask students to complete (Page 23) Task A, 'What is it used for?'. This task prompts students to consider which media they would use for a range of situations and why.	• Students will have already encountered the use of multimedia devices such as cameras in previous chapters, so this is a recap.
2	• Direct students to the 'Stay safe' message. Discuss responsible uses of cameras and sound recording devices with regard to online safety (Page 32). • Give students some time to think about what rules they should agree to before they begin to use the devices. As a class, agree a set of class guidelines for the use of multimedia devices. • Ask students to complete (Page 23) Task B, 'Class guidelines for safe use of multimedia devices'. Ask students to note down the finalised guidelines in their Workbooks.	• The exact guidelines you specify will depend on the rules within your school, and your specific requirements. The guidelines might include rules such as: • Do not video people without their permission. • You may not take recordings outside of the lesson. • Make sure everyone is happy with footage that is collected, and delete anything that makes someone unhappy.
3	• Demonstrate how to use the multimedia devices you have available in school.	• This section of the lesson will depend entirely on the devices you have available within school. Your school may have a 'bring your own device' policy that allows students to use their own devices for schoolwork. • You may want to enforce a rule that students are only video/sound recorded with their own device, so that they retain control over any recordings.

Activity	Instruction	Guidance
4	• Use (📖 Page 32) Build 3. Using whatever devices are available, students should work in pairs to record sound or video, or to take photographs to be used within the virtual tour. • If students are not creating a school tour, this might be a video interview about something in the tour, or a sound clip for a voiceover, for example.	• Careful classroom management might be necessary in order for all students to have the opportunity to successfully complete this task. Different groups may need to work in different areas of the room, or you may give permission for students to visit other areas outside the classroom. • If space is tight, perhaps have each pair of students group with a different pair, and they record each other. • If students are creating a virtual tour about a local landmark and it is not possible to go there during the lesson, they will need to problem solve, such as using Google Earth to work out how they might incorporate different media into their tour.
5	• Reflection: Ask students to reflect on how confident they felt about using the different devices? Then ask them to complete (📓 Page 24) 'Reflection'.	
	Homework (Optional) (📓 Page 24) • Students are provided with a variety of hypothetical scenarios where multimedia devices are used. Their task is to anticipate what could go wrong and identify the precautions they should take to make sure everyone stays safe.	• This task tests whether students can extrapolate from a scenario what the potential dangers might be. It might also test whether the class guidelines you came up with are sufficiently robust! • You could provide individual feedback to students by marking this task in their workbooks, or you could choose to go over it as a class during the following lesson in case there are any concerns.

Build on this: If you have the appropriate software within your school, and students have been able to record video footage, you could give them the opportunity to edit the footage. For example, they could cut parts out, add background music or add written titles. They may even be able to combine a recorded video with a separately recorded voiceover.

Chapter 2:	Lesson 5 Presentation software		
Student's Book page 33			
Learning Objective/s	• Use sound, video, text and other multimedia to create a digital artefact		
New keywords	**Hotspot** – an area on the screen that the user can click on to perform an action **Navigation** – the way a user moves through a presentation or a website		
You will need	• Presentation software	**Key pedagogy**	Develop practical skills

Activity	Instruction	Guidance
1	• Use Discuss 5 (Page 33). Using the screenshot and information (Page 33), ask students: 'Can you foresee any problems with putting navigation buttons in the document template?' • Ask students to complete (Page 25) Task A, 'Navigation'. Students draw a diagram to show how the different parts of their presentation will link together.	• Students may have already noticed at the stage they created their document template that there **may** be an issue with adding previous and next buttons to the template, because where they link to **may** be different on each slide. "Next" and "Previous" buttons are available to use, this is different to making a hotspot to link to somewhere else. • You can use the template to accurately position new navigation buttons on each slide, and then remove the buttons from the template afterwards. This will mean that the buttons always remain in the same place, keeping consistency between slides.
2	• Using the information provided (Page 34), demonstrate to students how to add a link to somewhere else in the presentation.	• You may need to also demonstrate how to order (layer) items on the page, so that a hotspot where users click on part of an image is not hidden behind the image itself. Most presentation software allows you to right click/control click on an element and select 'order' and then 'send backwards' or 'bring forwards' to change how it is layered. • You can use the analogy of different sheets of paper stacked on top of each other to explain this concept to students.

Activity	Instruction	Guidance
	• During Build 4, students work in pairs to put together their tours using presentation software (📖 Page 34).	• You may wish to give students advice or a reminder about working together on a project and check carefully that pairs are working collaboratively.
3	• Reflection: Ask students to reflect on working in pairs. Then ask them to complete (📖 Page 25) 'Reflection'.	• This reflection task provides an opportunity for students to reflect on their experience of working with a partner. This could be kept private to the student, or you could ask volunteers to share their thoughts with the class.
	Homework (Optional) (📖 Page 26) • Ask students to find out about accessibility, and research some advice for making their virtual tours accessible to as many people as possible.	• It is often interesting for students to consider accessibility, as it may not be something that comes up in everyday life. Often, accessibility adjustments, such as using clear fonts and sufficiently large buttons, can benefit all users, not just those with disabilities.
Build on this: Students could offer feedback to another pair, mid-way through the exercise.		

Chapter 2:	Lesson 6 Augmented reality		
Student's Book page 35			
Learning Objective/s	• List uses for augmented reality in familiar environments, including education and entertainment		
New keywords	**Augmented reality** – using technology to superimpose digital content on top of a real-world environment		
You will need	• Scissors (or pre-cut the worksheet) • 'Awards' worksheet	**Key pedagogy**	Showcase

Activity	Instruction	Guidance
1	• Run the showcase (📖 Page 35). Showcase 1. Each pair sets up their virtual tour on a device. Students circulate and try out other students' tours.	• If you prefer, you could ask one student from each pair to stay with their tour to answer any questions, while the other person circulates. Halfway through, they swap around so that everyone gets a chance to view the other tours.
2	• Hand out a copy of the 'Awards' worksheet to each pair. Once students have had a chance to try all of the tours, in their pairs they should decide who to give their awards to and write the reasons on the worksheet. • You can then collect in the finished worksheets and distribute the awards to the winners. You may wish to do this at the end of the lesson to ensure a smooth flow of progression.	• Either pre-cut the worksheet into individual awards, or if scissors are available students could cut their own. • Some pairs may not win any awards. You may wish to come up with your own teacher award using the blank certificate template on the worksheet, or you could ask each pair of students to come up with their own award title.
3	• Use Discuss 5 (📖 Page 35). Have a group discussion about augmented reality. What uses of augmented reality can students think of? Ask students to think about this question before they open their Student's Books. • Talk through some of the examples (📖 Pages 35 and 36).	• Students may be excited after the showcase part of the lesson, so this activity is an opportunity to transition between activities and settle them back down for a discussion.

Activity	Instruction	Guidance
4	• Ask students to complete (Page 27) Task A, 'What are the benefits and drawbacks of augmented reality?' Ask them to consider the uses presented in the Student's Book (Page 37), and note down any advantages and disadvantages.	• There are some benefits and drawbacks provided in the Student's Book) but this task asks students to draw on what they have learned in the lesson and their own understanding. • There are some suggestions of benefits and drawbacks provided in the answer document.

5	• Demonstrate an example of augmented reality and allow students to try it, if you have tablets or mobile devices available.	• Apps you could try:
		  • Measure app built in to iOS
		  • Adobe Aero
		  • BBC Civilisations AR
		  • WWF Free Rivers
		  • GeoGebra 3D Calculator
		• Apps may not be available in all countries or across all platforms. Search for each one using a search engine.

6	• Reflection: Ask students to reflect on their projects. Then ask them to complete (Page 28) 'Reflection'.	• You could ask students to refer back to this reflection before tackling a future project that involves formatting a document, for example Chapter 6. Can students act on any of their own advice?

	Homework (Optional) (Page 28)
	• Students consider how they would use augmented reality technology to enhance a *real-life* tour.
		• Ideas could include:
		  • virtual signposts to show visitors round a site
		  • visualisations of how a place may have previously looked, for example if the tour is of a ruined castle, augmented reality could show how it looked in the past
		  • overlays of activities taking place, for example to show what activities the place was used for in the past.

Build on this: Explore the difference between a tour that uses augmented reality and a virtual tour. Augmented reality superimposes extra features onto an existing environment, whereas virtual reality replaces the environment altogether. While students have not created a tour in this chapter using virtual reality, their tour is still virtual as it is possible for it to be experienced by people not in the same geographical location.

Chapter 3 Create with code 1

Project: Fantasy name generator

Chapter overview

Students may be familiar with small programs that generate fantasy names by joining together several randomly chosen parts. During this chapter they will learn how to use a program library to generate randomness in their program, before planning and writing code for a program to generate a fantasy name with multiple parts. They will add extra features to their program following an iterative process.

Chapter 3: Lesson summary		
Lesson	**Learning Objectives**	**Summary**
1	• Predict the outcome of algorithms and test that they meet the expected outcome	Students will familiarise themselves with how to use a Python development environment. They will compare Scratch code with Python code for variables, input and output. Then, they will practise these skills to produce a funny story game.
2	• Describe the purpose of program libraries • Use program library functions in text-based programs	Students will learn about program libraries in general, and specifically how to use Python's random library. They will use the library to choose random integers and random items from a list and then practise these skills by writing a password generator program.
3	• Decompose a problem into sub-problems	Students will practise decomposing the task of creating a fantasy name program into smaller parts. They will then plan each part using a flowchart.
4	• Develop an algorithm that uses a constant	Students will learn about constants and where they can be used. They will then learn about comments and use them to form a comment skeleton based on their flowchart plans. This will then help students write the code for their name generator programs.
5	• Develop a program using an iterative process	Students will continue to build upon their programs by adding extra features. They will learn some debugging skills specific to Python and use these when writing their code.
6	• Develop a program using an iterative process	Each student will showcase the program they have written to another student, who will perform a code review. They will then give user feedback on a program. Students then use the feedback they have received to improve their programs.

End of chapter project	Fantasy name generator
Example ideas that learners could design:	• A fantasy story about a hero who goes on a quest • A detective story where the hero solves crimes

Why we are learning this	
Useful skills covered in this chapter	• Providing useful feedback to peers • Decomposition – breaking down a difficult task into smaller parts
Links to future careers	• Software developer – students will practise performing a code review • Author – creating a story and character names
Links to other subjects	• English – writing stories

Chapter 3 – Create with code 1

Project: Fantasy name generator

Name	

	Working towards (1)	Meets expected (2)	Exceeding expected (3)	Score
Planning	• Shows evidence of decomposing the task of creating a fantasy story program into smaller parts	• Plans each part of the program using a flowchart	• Shows effective use of comments to improve their design based on their flowchart plan	
Program design	• Writes the code for generating a simple fantasy story	• Uses Python's random library effectively to create a program that generates a fantasy story	• Uses constants and variables appropriately in their program	
Program development	• Develops their program by adding extra features as part of an iterative process	• Demonstrates some debugging skills specific to Python and shows evidence of using these to improve their code	• Shows evidence of improving their own program based on feedback received	

Teacher feedback	
Learner response	

Chapter 3: Lesson 1 Familiarisation with Python

Student's Book page 40

Learning Objective/s	• Predict the outcome of algorithms and test that they meet the expected outcome		
New keywords	**Concatenation** – joining two or more pieces of text together **Text-based language** – allows you to type text instructions that a computer can run **String** – a sequence of characters (letters, numbers and/or symbols/punctuation marks), usually written within quotation marks		
You will need	• Devices capable of running Python • Scrap paper • Pencils • A Python coding environment	**Key pedagogy**	Develop practical skills

Activity	Instruction	Guidance
1	• Demonstrate and run a working name generator program you have written. • Note that each time you run the program, a different name is generated – this is because the choice is random.	• An exemplar program you could demonstrate is provided in the answer section for Lesson 5.

Activity	Instruction	Guidance

2	• Tell students how they will access Python (this will be specific to your school).
	• Use Build 1 (Page 40) In the Build task, students will test that Python is working by running the 'Hello world' program.
	• Ask students to complete (Page 30) Task A, 'Writing a Python program'. Students note down instructions so that they can remember how to access Python in future lessons.
	• Use Discuss 1 (Page 41). Did the program do what you expected it to? What do you think the command `print` means in Python?
		• How you run Python in your school will depend on what technology you have available and whether you have reliable internet access. Some options include:
		  • Installing Python IDLE on your school network
		  • Using Python online in a browser, for example trinket.io, repl.it, EduBlocks
		  • Using Python on a Raspberry Pi
		  • Using an app to write Python code on a tablet such as an iPad.
		• You will need to research and test what is available to you in advance of this chapter, so that you can advise students.
		• Remember to demonstrate how to open the development environment, type in the environment, save and then run the program. Students will note these instructions down in the workbook task to refer to if they forget how to do this later in the chapter.
		• Students sometimes expect the `print` command to print the message onto paper using a printer.

3	• Show the comparison of Scratch and Python inputs (Page 41). • Use Build 2 (Page 41). Students recreate the example. • Ask students to complete (Page 30) Task B, 'Recreate the Scratch program'. Students write a program to recreate a Scratch program with an input and output, then test whether they are right.	• Students sometimes get confused when a Python program stops after an input, e.g. "What is your name?" in the example. Explain that Python is waiting for you to type something and then press 'Enter' on the keyboard, just as in Scratch you would type something and then click the tick icon. Use the Scratch screenshot (Page 41) to help with the explanation, or a live demo of you typing a Python program.

Activity	Instruction	Guidance
4	• Ask: 'What if we want to save more than one input?' Answer: You can do this using a variable for every additional input. • Show the updated Scratch program (Page 42) with the Python equivalent. • Use Build 3 (Page 42). Ask students to adapt their programs to include at least two variables populated via input, and two statements that repeat back the data stored in the variable with an accompanying comment.	
5	• Organise students into groups of four, and hand each group a piece of scrap paper. • One person secretly writes down the name of a **person** at the top of the paper, and then folds the top of the paper down so the other group members cannot see it. • They then hand the paper to the next person who writes down a **place**, then the next writes a **vegetable** and the next writes down an **emotion**. • Then, reveal a story with blank words that you have written, and ask students to unfold the paper and read the story, substituting their words into the blank spaces.	• An example of a story might be: One day, <person> went to <place>. As soon as they got there, they slipped on a <vegetable>! This made them feel very <emotion>. • You can adapt the story to be relevant to your school or context as appropriate for your learners.
6	• Use Build 4 (Page 43). Students design and write a program that plays the funny story game.	• Students should use a maximum of four inputs so that they are able to spend sufficient time on all parts of the program, otherwise they may focus on collecting a lot of input and then not finish the output part of the program.
7	• Reflection: Ask students to complete (Page 31) 'Reflection'. What differences did they spot between Python and Scratch? Which do they prefer, and why? • Use Discuss 2 (Page 43). Ask students to share any similarities and differences they noted with the class.	• Students will often prefer the language they are more familiar with (i.e. Scratch) at this stage. This lesson uses lots of comparison between the two because transitioning between a block-based and a text-based language can be tough for students as it will initially slow them down as they need to remember syntax.
	Homework (Optional) (Page 31) • Write down the output of each of the programs.	• Tests the skills learned in the lesson

Build on this:

What happens if you try to input a number? For example:

```
age = input("How many years old are you? ")
```

If the user types in 12, you can concatenate this into a string just like any of the other inputs we have seen. However, if you try to do calculations with it, something unexpected happens!

```
age = input("How many years old are you? ")
double_age = age * 2
print("If you were twice as old, you would be " + double_age)
```

You will get this output:

How many years old are you? 12
If you were twice as old, you would be 1212

Why do you think this is?

Answer: When you use the Python command input() to get the user to type something in, the input is always saved as a string. If you try to multiply any string by 2, you will get the same string repeated! For example: `"abc" * 2 = "abcabc"`.

When the user types in a number, you must first convert it from a string to a number to be able to use it in calculations. In this code, the variable age has been turned into an integer before being multiplied by 2. In programming, this is called casting – changing the data type of a piece of data.

```
age = input("How many years old are you? ")
double_age = int(age) * 2
```

Chapter 3: Lesson 2 Program library

Student's Book page 44			
Learning Objective/s	• Describe the purpose of program libraries • Use program library functions in text-based programs		
New keywords	**List** – a data structure that can hold more than one piece of data **Program library** – a collection of pre-written code that can be imported and used within another program **Random** – chosen by chance		
You will need	• Devices capable of running Python	**Key pedagogy**	Lead with concepts

Activity	Instruction	Guidance

1	• Introduce the concept of a program library – a collection of pre-written code that can be imported and used within another program.
	• Ask students to complete (Page 32) Task A, 'Video game program library'. In pairs, students write down a list of functions that might be in a video game library. Students could use the picture of a video game (Page 44) for inspiration.
	• Use Discuss 3. Ask students to share their ideas with the class (Page 44).
		• The answers that students generate could be varied. In general, the functions should be relevant to video games, self-contained and likely to be common to many games.
		• Good examples:
		  • Make a character jump
		  • Collect an item
		  • Display a menu screen
		• Poor examples:
		  • Send an email (not relevant)
		  • Create a level (scope too large)
		  • Do a super triple flaming spike jump (too specific to one game)

2	• What are the benefits of using a library? Talk through the benefits (Page 45).	• Students may be able to think of some more benefits at the end of the lesson once they have used a library.
3	• In this chapter, students will use the random library. • Ask students to complete (Page 32) Task B, 'Guess the output'. Run the sample program and predict the output. • Discuss: Did everyone get the same result? Why not?	• The output of the sample program will be different every time the program is run, because the program will generate a random (integer) number between 1 and 6. • For any students who are finished quickly, you could ask them to investigate what happens if they change the numbers 1 and 6. • Answer: These are the bounds for the random number.
4	• Show a new function from the random library – choosing a random item from a list (Page 46). • Use Build 5 (Page 46). In the Build task, students type in and run the code to select a random item from a list.	• The list data type is only briefly explained but it should be sufficiently clear for students to use it in a limited way as in the example. They will not be expected to append or remove items from a list or access them by index in this chapter.

Activity	Instruction	Guidance

5	• Use Build 6 (Page 47). In the Build task, students design and build a password generator program to generate a secure password. The password should include :
	  • a number
	  • a letter
	  • a special character
		• At this stage, students have not learned about loops in Python so you should expect that they will generate each character for the password individually.
		• Students may discover that a random choice is chosen at the time that line of the program is run, and it does not change. For example, they may expect this code to generate two different special characters.
		`special = random.choice(["*", "%", "!", "?"])`
		`print( special + special )`
		• The program is executed line by line, so the value of the variable `special` is assigned to one of the values from the list, before the second line of code is executed. The result will always be two of the *same* character, although the character will be chosen at random each time the program is run.
		• A sample answer for this task is provided in the answer document.

6	• Reflection: Ask students to complete (Page 33) 'Reflection'. This asks students to reflect on the techniques they have learned during this lesson.	
	Homework (Optional) (Page 33) • Describe the purpose of a program library and show understanding of how two of its functions work.	

Build on this: Discuss: When you wrote the first program to generate a random number between 1 and 6, the output was different every time you ran the program. But, given that you know a computer can only perform specific instructions, how can a computer select a truly random number?

Answer: It cannot! The numbers generated by the computer are 'pseudo-random', because they are based on a starting number called a seed, which changes its value (based on the computer's current time) every time the program runs. If you want to prove this to yourself, run the following program, which fixes a specific seed. You will get the same three "random" numbers every time you run the program. Change the seed value to a different number and you will get a different set of three "random" numbers each time you run the program.

import random

```
random.seed(90)
print(random.randint(1, 6))
print(random.randint(1, 6))
print(random.randint(1, 6))
```

Chapter 3:	Lesson 3 Decomposition
Student's Book page 48	
Learning Objective/s	• Decompose a problem into sub-problems
New keywords	**Decomposition** – the process of breaking down a complex problem into smaller, more manageable tasks
You will need	• A3 or A4 paper • Pencils

Activity	Instruction	Guidance
1	• Introduce decomposition via the analogy of learning something new (Page 48). • You wouldn't try to learn the whole thing at once, instead your teacher or coach might break it down into individual skills.	• If there is a particular sport that you know lots of your students play, you could use that as an example instead of football.
2	• Ask students to complete (Page 34) Task A, 'Decompose a game'. Students practise decomposing a game into smaller subtasks	• You may want to specify a game to decompose so that students can compare results afterwards.
3	• Ask students to complete (Page 34) Task B, What's in a name? This asks students to look at an example of a fantasy name and decompose it into smaller parts	• Students should describe the component parts of the names they see. For example, the first word in each name is an adjective (describing word) - fearless, cheerful etc.

4	• Go over the example flowchart (Page 50) for generating the adjective part.
	• Explain that in Design 1, students will draw three more flowcharts to plan the three remaining parts of the name generator.
		• You may want to do a quick recap of the meanings of the flowchart symbols:
		  • rounded rectangle – start or end of the flowchart (terminator)
		  • rectangle – process (including assigning a value to a variable)
		  • parallelogram – input or output (e.g. user typing in, or printing a value to the screen)

5	• An important part of decomposition is being able to put the component parts back together again to create the whole program. • In Design 2 (Page 50) students create a single flowchart of the full program design, including how the four parts of the name will be output at the end.	• Students may find that they run into problems when putting the decomposed parts back together. For example, if they have assigned all of the parts of the name to a variable called 'word' in their individual flowcharts, they will need to redraft and name each one differently in the main flowchart.
6	• Reflection: Ask students to complete (Page 35) 'Reflection'. Did they find decomposition helped them?	
	Homework (Optional) (Page 35) • Students are asked to decompose instructions for drawing something.	• You could ask students to swap books at the start of the following lesson and follow each other's decomposed steps – how good are their instructions?

<table>
<tr><th>Activity</th><th>Instruction</th><th>Guidance</th></tr>
<tr><td colspan="3">

Build on this: If you wish, you could introduce the flowchart symbol for a subroutine, which is a rectangle with an extra line on each side.

Figure 3.3.1 Flowchart subroutine symbol.

A single subroutine symbol can be used to represent each of the four individual flowcharts. This is a useful shorthand to avoid unnecessary copying of previous work.

Students will *not* be expected to use subroutines in their code at this stage.

</td></tr>
</table>

Chapter 3:	**Lesson 4 Name generator**
Student's Book page 51	
Learning Objective/s	• Develop an algorithm that uses a constant
New keywords	**Constant** – a value that does not change while the program is running **Comment** – a note written for a human to read in a program. All comments are completely ignored by the computer
You will need	• Devices capable of running Python • Flowchart plans from the previous lesson

Activity	Instruction	Guidance
1	• Recap the code for choosing a random word from a list (learned in Lesson 2) (Page 46). • Use Discuss 4 (Page 51). Which variable's content changes during the program? Which stays the same? Talk this through with the class.	• The variable 'seasons' will stay the same – the items in the list do not change. The variable 'result' will be different each time the program runs, because it is randomly assigned one of the seasons from the list.

2	• Talk through the two pieces of code for calculating the area and amount of money saved when ordering a pizza (Page 51).
	• Use Discuss 5 (Page 51). Ask students to discuss in pairs which code is clearer, and why?
	• Ask students to complete (Page 36) Task A, 'Constants'. This asks students to complete the sentences to explain some of the benefits of using constants.
	• In Design 3, ask students to mark on their flowcharts which pieces of data should be stored as a variable, and which should be stored as a constant (Page 52).
		• Avoid defining a constant as 'a variable which does not change'. It is confusing to use the word variable in the definition of a constant, because a variable, by its very definition, varies!
		• This is a complicated program to show students at this stage, but this is deliberate because the use of constants should make it easier to understand.
		• Constants:
		  • make the code easier for other people to understand
		  • are easy to find and change within the program, e.g. if the discount changes you can update it once rather than having to go through and find everywhere in the program you have written 0.1.
		• Make it unlikely you will accidentally change their value during the program by writing the name in capitals.

3	• Introduce comments – they are there for the programmer to read but are completely ignored by the program when it is run. • In Build 7, students start a new Python program. They need to go through the flowcharts they drew in the previous lesson and add one comment to the program for each box in the flowchart (Page 52).	• This method may help students to translate their algorithm from a flowchart into real Python code. It is sometimes called writing a comment skeleton. • Comments can be used as a tool both tool, either to note down a task that needs to be done *before* the code is written or to explain code *after* it has been written. • Other programming languages enforce the rule that a constant cannot be changed while the program is running. Python does not enforce this rule, so it is technically possible – but inadvisable – to change the value of a constant in a program.

Activity	Instruction	Guidance
4	• Students write the name generator program in Build 8. (📖 Page 52). Students write program code underneath each comment to complete the task described. This will build up to a finished program.	• Comments are usually written either on the line above the code they refer to or on the same line. For example, this is a valid comment: `print("hi")   # Say hi` • The # tells the computer to ignore everything from the # symbol to the end of that line. • Make sure students save their code regularly. Demonstrate again how to do this if you think it is necessary.
5	• Reflection: Ask students to complete (📖 Page 36) 'Reflection'. Students reflect on what advice they would have given themselves at the start of the lesson.	
	Homework (Optional) (📖 Page 37) • Students are asked to add comments to the code to explain what is happening.	

Build on this: If you need to write a comment which needs more than one line, either start each line with a #, or use triple quotation marks to start and end a very long comment.

```
"""
Calculate the area of the pizza
The formula for the area of a circle is Pi * R squared
"""
area = radius * radius * PI
```

Is it a good idea to always comment every line of code in your program?
Answer: Not every line in a program needs to be commented. There is no benefit to commenting a line of code where the meaning of the code is obvious, for example:

```
# Print a message saying Hello world
print("Hello world")
```

However, most students do not comment their code enough! A good incentive for them to comment might be that it is much easier for someone else to help them debug the code if they run into problems if there are sensible and helpful comments.

Chapter 3:	Lesson 5 Story generator		
Student's Book page 53			
Learning Objective/s	• Develop a program using an iterative process		
You will need	• Devices capable of running Python	**Key pedagogy**	Develop practical skills

Activity	Instruction	Guidance
1	• In this lesson, students will add a story for their character. This will be similar to the one from Lesson 1, but parts of the story will be randomly generated instead of typed in. • Demonstrate an example you have written in code or refer to the code in the Student's Book (Page 53). • Ask students to complete (Page 38) Task A, 'Design your story'. Students need to think of some ideas for random choices to help to tell a story.	
2	• Ask students to complete (Page 38) Task B, 'What does the error message mean?'. Ask students to complete this task and see if they can work out what each error message means. • Go over the answers in class, and ask students to correct or write in any answers they did not get right.	• Testing as they go means that students can narrow (running and checking code) down which sections of code may have bugs. • Being able to understand error messages makes students more self-sufficient at debugging their own code. • Sometimes an error message will give a line number, but the error it refers to is on the previous line. For example, if a student forgets to close a bracket, Python will keep looking for the closing bracket on the following line of code and will only then give an error when it doesn't find one.

Activity	Instruction	Guidance
3	• Remind students that they may want to use the comment skeleton technique to add new parts to the program. • Remind students that each time they add a new part of the story, they should stop and run their program to check it still works. • Use Build 9 (Page 54) Ask students to save a copy of their name generator programs and add a story.	• Students will work at their own pace, and there will be differentiation by outcome. • Using conditional statements in Python is covered in Chapter 8.5. However, some more advanced students may feel frustrated that they don't yet know how to use this construct, which they are comfortable using in a block-based language such as Scratch. You could suggest that these students research how to use an 'if' statement in Python (either online or via the materials in this book) and attempt to add a condition to their story, as an extension task.
4	• Reflection: 'Ask students to reflect on whether there is anything they would like to add to their program but may not know how to program? Then ask them to complete (Page 39) 'Reflection'.	• Some students may be feeling frustrated at this point that there are things they would have liked to do in their story and could have done in another language such as Scratch, but they don't yet know how to do them in Python. For example, making a decision in the story. • You can reassure students that they will learn all three constructs in Python.
	Homework (Optional) (Page 39) • This asks students to debug the code.	• Research shows that being able to read and interpret code is very important.

Build on this: Using randomness in a story may result in scenarios that don't quite make sense. Look at these two lists:

```
JOBS = ["musician", "farmer", "cleaner"]
ACTIONS = ["plays music", "keeps animals", "cleans houses"]
```

If you choose one item at random from each of the lists, your character may end up being a musician who keeps animals, or a farmer who plays music. Instead, you may want to randomly choose a job, but then use the corresponding action in your story.

Each item in a list has a number (called an index), and you need to be careful because the numbers start at 0, i.e. musician is item 0 in the list. So, you could generate a random number instead of a random choice, and then use it to reference the item at a specific position in the list, like this:

```
import random
name = input("Please type in your name: ")
JOBS = ["musician", "farmer", "cleaner"]
ACTIONS = ["plays music", "keeps animals", "cleans houses"]
num = random.randint(0,2)     # Choose a random number between 0 and 2
job = JOBS[num]               # Select the job at this index
action = ACTIONS[num]         # Select the action at this index
print(name + " is a " + job + " who " + action)
```

Assuming the items in your two lists are in the corresponding order, this means that you can guarantee to choose a random job from the JOBS list, and always get the correct corresponding action from the ACTIONS list.

Chapter 3: Lesson 6 Feedback on software

Student's Book page 55	
Learning Objective/s	• Develop a program using an iterative process
New keywords	**Code review** – a software developer checking code written by another developer and suggesting improvements **Iterative development** – regularly adding new features and improvements to a piece of software, and then seeking feedback to inform the next set of improvements **User feedback** – feedback from a person who uses a piece of software
You will need	Devices capable of running Python

Activity	Instruction	Guidance
1	• Students will take on two different roles – software developer and software user – to evaluate a partner's work so far.	
2	• Use Discuss 6 (Page 55). As a class, discuss the examples and decide which features of the code will form part of the code review, then note these down in (Page 40) Task A, 'Code review focus'. • Ask students to complete (Page 40) Task B, Review the code. In pairs, students swap computers and code review their partner's code. (Students write comments in their partner's Workbook.)	• You could guide the code review so that it focuses on particular areas you want students to focus on, or you could just use the four areas listed in the Student's Book. • Students should write in their **partner's** Workbook so that each student has a note of the feedback that is relevant to them. • You may need to remind students to write neatly and carefully as they are writing in each other's books.
3	• Introduce the definition of 'user feedback' and use Discuss 7 (Page 56). As a class, discuss and decide which features of the program you will provide user feedback on, then note these down in (Page 41): Task C 'User feedback focus'. • Ask students to complete (Page 41): Task D, 'Give user feedback'. In pairs, students swap computers and provide user feedback. (Students write comments in their partner's Workbook.	• Students should write in their **partner's** Workbook, so that each student has a note of the feedback that is relevant to them.
4	• In Build 10 (Page 56) students return to their own computers and look at the feedback they have been given. They then have some time to change their code and implement any feedback.	• Ensure that you leave enough time for this activity and do not spend too long on the reviewing.
5	• Reflection: Ask students to complete (Page 42) 'Reflection'. This task asks students to reflect on the quality of feedback offered by peers.	• Encourage students to be honest (but polite) in their Workbooks as this is a reflection task. It is not intended to be shared back to the person who wrote the feedback; it is intended to make the student think about what makes good feedback.
	Homework (Optional) (Page 42) • Students are asked to think in more depth about some of the terminology they have met during this lesson.	• There is some supporting information in the Student's Book (Pages 55 and 56), but this task is intended to make students think and suggest reasons, rather than list facts.

Build on this: You could build code reviewing and user feedback into future programming lessons. You may even want to try out **pair programming**, which is a technique commonly used in industry where two programmers work together at one workstation. One is the 'driver' who writes the code, and the other reviews each line as it is typed. They switch roles regularly, to avoid the reviewer becoming disengaged.

This may also be a good strategy if you have students at different levels of confidence with programming but ensure that all students are engaged and roles are regularly switched to avoid one student doing the majority of the work.

Chapter 4 How computers work

Project: Troubleshooter

Chapter overview

Just like a doctor learns about human biology so they can diagnose diseases, if you know how your computer works, you can use this information to sort out any problems. In this chapter students will learn about various parts of the computer, and then combine this knowledge to create a troubleshooter that provides advice to help with a computer problem.

Chapter 4: Lesson summary		
Lesson	**Learning Objectives**	**Summary**
1	• Describe the roles of main memory, RAM, and ROM	Students are introduced to the project – to create a troubleshooter. They think about the differences between RAM and ROM and find out how RAM affects the computer's performance.
2	• Describe the purpose of an operating system	Students investigate the role of the operating system, including scheduling and memory management.
3	• Describe why utility programs are needed • Explain the role of antivirus and antispyware on a network	Students find out about a range of utility programs. Then they play the role of a virus trying to get past antivirus software on a network.
4	• Explain the role of a firewall • Explain how file access permissions are used within a network	Students are shown how to protect a network, both from external threats with a firewall, and internally using file access permissions.
5	• Describe the advantages and disadvantages of wired and wireless networks, including performance and security • List the benefits and risks of the Internet of Things (IoT)	Students compare wired and wireless networks and discuss their advantages and disadvantages. They think about the risks of IoT devices connecting to a network and whether the benefits outweigh these risks.
6	• Follow conditional statements in flowcharts	Students role-play a user and a technician, and use the troubleshooter they have developed throughout this chapter to solve problems.

End of chapter project	Troubleshooter
Example ideas that learners could design:	• Troubleshooter to find and resolve a range of computer problems

Why we are learning this	
Useful skills covered in this chapter	• Learners will be able to understand the cause of and be able to resolve a range of basic computer issues
Links to future careers	• General ability to solve IT problems in any career • Technical support skills
Links to other subjects	• Photography and media – things to consider when editing large image or video files • Biology – classification of animals

Chapter 4 – How computers work

Project: Troubleshooter

Name	

	Working towards (1)	Meets expected (2)	Exceeding expected (3)	Score
Prevention of computer problems	• Understands why utility programs are needed	• Understands the role of antivirus and antispyware on a network and the risks of IoT devices connecting to a network	• Understands how to protect a network, both from external threats with a firewall, and internally using file access permissions	
Identify common problems	• Shows understanding of some common computer problems	• Shows understanding of computer problems that might relate to tasks handled by the operating system	• Correctly identifies which tasks are handled by the operating system and which would be handled by application software	
Create a troubleshooter	• Uses conditional statements in flowcharts to develop a simple troubleshooter program that can diagnose and solve a limited number of computer problems	• Uses conditional statements to develop a simple troubleshooter program that can diagnose and resolve a range of basic computer issues correctly	• Uses conditional statements to develop a troubleshooter program that can diagnose and solve a wide range of problems	

Teacher feedback	
Learner response	

Chapter 4:	Lesson 1 Memory		
Student's Book page 59			
Learning Objective/s	• Describe the roles of main memory, RAM, and ROM		
New keywords	**Troubleshooter** – a tool or guide that gathers information to pinpoint the cause of a problem, and then offers advice on how to fix it **RAM** – Random Access Memory, used to store instructions and data the computer is currently working with. Contents are lost when the power is switched off **ROM** – Read Only Memory, stores the instructions needed to load the operating system. Contents are kept when the power is switched off		
You will need	• Coloured pencils or pens • Flowchart software (see recommendations in 'Start your troubleshooter' section below) • Mini whiteboards or cards (optional)	**Key pedagogy**	Unplugged

Activity	Instruction	Guidance
1	• Introduce the lesson with a short task – classify the animals using the flowchart. • Ask students to complete (Page 44) Task A, 'Classify the animals'. This asks students to follow the flowchart to classify four animals according to whether or not they are warm-blooded and have a backbone.	• Encourage students to follow the flowchart in order from the start, otherwise they may mis-classify snake and worm.
2	• Use Discuss 1 (Page 59) Ask 'Where have you seen a troubleshooter before? '	• Examples are provided in the Student's Book. • If there is one you are familiar with, you could demonstrate an example of a troubleshooter, for example a medical diagnostic tool.

Activity	Instruction	Guidance
3	• Introduce ROM and RAM and talk through the information on (📖 Pages 60 and 61). • Ask students to carry out Role play 1 (📖 Page 61) Students will work in pairs, and will need five coloured pencils plus the colouring page (📓 Page 45): Task B, 'Memory role'. Student A should sit a metre away from student B. • Explain that student A holds all of the pencils – this is the **secondary storage**. Provide a copy of (📓 Page 45) for student A. • Student B does the colouring, this represents the **processor**. This student should begin the exercise by filling in the key at the top of their Workbook page with each of the five colours. • Student B can also keep a maximum of two pencils in front of them on their desk – this represents the **RAM**.	• Student A calls out the instructions in order to student B, who colours the picture in their Workbook. • If RAM is full, one of the slots must be emptied to be able to take a new colour. • If a colour that the processor does not have in RAM is required, it must be requested from secondary storage, and it must take up one of their two RAM slots. Student B must get up and retrieve the pencil from student A.
4	• Students swap roles. Increase the amount of 'RAM' available in the role-play exercise by allowing a maximum of four pencils on the desk for this round, then run the activity again. • (📖 Page 61)	• Students swap roles and perform the exercise again but with the increased RAM availability. • See answer document for answers to discussion questions.

<table>
<tr><th>Activity</th><th>Instruction</th><th>Guidance</th></tr>
<tr><td>5</td><td><ul><li>Use Build 1 (Page 62). In the Build task, students start to create a troubleshooter flowchart with flowchart software, by adding questions relating to memory, and solutions. For example:</li></ul>Does your computer start up? YES →
Is your computer running slowly? YES →
Are you running lots of programs at once? YES
→ Your computer is using all of its RAM. Either close some programs or upgrade your computer to increase the amount of RAM.</td><td><ul><li>There are lots of types of flowchart software available, and many are free (although some require an account). Examples include:<ul><li>Draw.io (free, can save to device without an account)</li><li>Microsoft Visio (paid)</li><li>Miro (free, needs account)</li><li>SmartDraw (free)</li></ul></li><li>It is OK if students spend some time familiarising themselves with the software during this lesson and don't add lots of questions, as there will be time during other lessons too.</li><li>Remember to explain to students how they save their work on the specific software you are using, so that they can access it again to work on during the next lesson.</li></ul></td></tr>
<tr><td>6</td><td><ul><li>Reflection: Ask students to complete (Page 46) 'Reflection'. Students reflect on whether there have been any problems, and how they might resolve these in the next lesson.</li></ul></td><td><ul><li>If you have noticed that several students shared a common problem, you could initiate a class discussion and encourage students to offer each other advice on how to resolve their issues.</li></ul></td></tr>
<tr><td></td><td>Homework (Optional) (Page 46)<ul><li>Students are asked to find out the specification of a particular computer and compare RAM to secondary storage.</li></ul></td><td><ul><li>This could be for a computer the student owns, or a 'dream' computer they have researched.</li></ul></td></tr>
</table>

Build on this: Have a RAM, ROM or secondary storage quiz. Call out some statements and students have to hold up a card or mini whiteboard with whether the statement applies to RAM, ROM or secondary storage. Some of the statements may apply to more than one. Here are some examples:

- Keeps its contents when the power is switched off (ROM, secondary storage)
- Used for storing files (secondary storage)
- Stores the instructions to boot the computer (ROM)
- Stores instructions and data the computer is currently working on (RAM)
- Relatively small capacity (ROM, RAM)
- Large capacity (secondary storage)

Chapter 4:	Lesson 2 The operating system
Student's Book page 63	
Learning Objective/s	• Describe the purpose of an operating system
New keywords	**Application software** – allows the computer user to do a task, for example a web browser or a word processor **Driver** – a small piece of software that allows the operating system to communicate with a hardware device such as a printer **Operating system (OS)** – an important piece of software that manages all other software and hardware devices **Utility software** – keeps the computer secure and in working order, for example antivirus software
You will need	• Flowchart software

Activity	Instruction	Guidance
1	• Talk through the information about the operating system and its role (Pages 63 and 64). • Ask students to complete (Page 47) Task A, 'Which tasks are performed by the operating system?'. This task asks students to identify whether or not a selection of tasks are carried out by the OS	• If you prefer, you could run this exercise as a card sort activity, with students working in pairs. You could print out the statements on paper and cut them into individual cards to be categorised. • Utility software is commonly considered to be part of the operating system, as it often comes as part of the OS software. It is covered in more depth in Chapter 4.3.

Activity	Instruction	Guidance

2	• Use Discuss 3 (Page 65) Ask the whole class to discuss whether they can come up with any computer problems that might relate to tasks handled by the operating system. Note them down on the board as inspiration for questions students might want to add to their troubleshooting tool.
	• Ask students to complete (Page 47) Task B, 'Troubleshooter questions'. Students are asked to plan some questions based on tasks performed by the operating system or utility software.
	• Use Build 2 (Page 65) where students add the questions they planned to their troubleshooter flowchart from the previous lesson.
		• Questions students may add as a result of this lesson could include:
		  • Is your computer starting up?
		  • Is your printer working correctly?
		  • Can you find the file you need?
		  • Is your keyboard working?
		• Remind students that it is not only important to ask questions in their troubleshooting tool, but also to diagnose the actual problem and offer advice for how to resolve it.

3	• Reflection: Ask students to complete (Page 48) 'Reflection''. How confident are students?	• You could invite students to share anything they are unsure about and ask other members of the class to put their hands up if they were also wondering the same thing. Another student who is confident with the concept in question may be able to support classmates with an explanation.
	Homework (Optional) (Page 48) • Students are given an image and asked to identify which tasks are handled by the operating system, and which would be handled by application software.	

Build on this: Students may discover that they do not know how to resolve some of the computer problems the class thought of during the discussion. As an extra activity in the final part of the lesson, you could divide up the problems from the 'Discuss' task and ask different groups to research the answers and share them back with the class.

Chapter 4:	Lesson 3 Utility programs		
Student's Book, page 65			
Learning Objective/s	• Describe why utility programs are needed • Explain the role of antivirus and antispyware on a network		
New keywords	**Compression** – significantly reducing the amount of storage space a file requires **Malware – mal**icious soft**ware**. Any software that intends to cause harm to a computer, for example a virus		
You will need	• No special equipment	**Key pedagogy**	Real-world contexts

Activity	Instruction	Guidance
1	• In the previous lesson students encountered the definition of a utility program. It would be a good idea to recap this. • Use Discuss 4 (📖 Page 66) • Go over the examples (📖 Page 66).	• There are other types of utility program not mentioned here. Broadly speaking, a utility program is any program that performs a maintenance or security task on the computer. • Other examples include a disk defragmenter (though most modern computers use solid state drives, which do not need to be defragmented), and system manager, which shows information about the system and its resources.
2	• (📖✏ Page 49) Task A, 'Purpose of utility programs'. Students are asked to match each task to the correct utility program.	
3	• Go over the material about anti-malware software on a network, including the case study (📖 Page 67). • Ask students to complete (📖✏ Page 49) Task B. Students pretend that they are network IT technicians at a large company and are giving advice to various employees.	• You could go over the answers to the Workbook task as a whole class activity, or you could ask students to compare answers in pairs or groups.

Activity	Instruction	Guidance

4	• Introduce Build 3 (Page 67). Ask students to add questions to their troubleshooter tools about utility software.	• Some example questions are:
		  • Can you find the file you need?
		  • Are you able to unzip the file?
		  • Is there a ransom message?
		  • Is the computer behaving strangely?
		  • Has anyone told you they received a strange email from you?
		• There is deliberately more time allocated to this task during this lesson, as previous lessons have had more new content to learn. Now that students are equipped with three lessons worth of learning, they can cover a variety of scenarios in their flowcharts.

5	• Reflection: Ask students to complete (Page 50) 'Reflection'. This asks students to reflect on how well the process of creating the troubleshooter is going, and whether there are any pathways for which they don't have particular answers.	• Some students may find it easier to create a separate troubleshooter flowchart for each lesson, as the flowchart can get quite large.
	Homework (Optional) (Page 50) • Students create a poster to put on the classroom wall, advising people about how to protect their computers from malware.	

Build on this: Research a specific case study about a piece of malware. What was the malware called, what type of malware was it, and what was the impact? How would you either prevent it from infecting your computer, or resolve the problem if you were infected?

Chapter 4:	Lesson 4 Unauthorised access		
Student's Book page 68			
Learning Objective/s	• Explain the role of a firewall • Explain how file access permissions are used within a network		
New keywords	**File access permissions** – restrict access to files and folders on a network to only the people who need to see them **Firewall** – provides a layer of protection between two networks, preventing threats from entering, and sensitive information from leaving **Port** – a number assigned to each type of network traffic going in or out of a firewall		
You will need	• 'Firewall rules' worksheet	**Key pedagogy**	Collaboration / Unplugged

Activity	Instruction	Guidance
1	• Introduce the definition of 'firewall' and talk through the diagram (📖 Page 68).	• Firewalls prevent unauthorised traffic reaching the network and prevent unauthorised traffic going out. • Students are likely to have encountered a firewall if they have tried to access a site that is not permitted by the school.

Activity	Instruction	Guidance
2	• Divide students into groups of two or three and investigate how a firewall works (📖 Page 69). • Provide each group with a copy of 'Firewall rules' worksheet, which lists the rules to specify which traffic is and isn't allowed through the firewall. • Students work together collaboratively to decide what happens in the case of each of the requests and note down the resulting decision (📖✏ Pages 51 and 52) Task A, 'Simulate a firewall'. • Go over the results with the class once students have completed the task.	• This activity is a simplified representation of how a firewall works. Each connection would have both an outbound and an inbound port number, which might be different, but this has been simplified to just one port number for the example. • Default port numbers are mostly used in the exercise, but students are not expected to know these. For example, HTTP traffic (unencrypted web traffic) would be port 80, and HTTPS would be 443. Port 22 is used for SSH (secure shell), which is used to connect remotely to computers – an example of this might be connecting to a company web server to change some settings. • Most port numbers are allocated to a specific type of traffic. Many firewalls operate an 'allow list' policy – this means that all traffic is blocked by default unless it is specified as being allowed. The alternative would be a 'block list' policy where all traffic is permitted unless specifically blocked. You may have heard these terms referred to as 'whitelist' and 'blacklist' respectively, but this terminology is outdated and should not be used.

Activity	Instruction	Guidance
3	• Firewalls are only effective at preventing external threats from reaching the network. They cannot protect against internal threats. • Talk through the material in about file access permissions (Page 70). • Use Discuss 5 (Page 71). Read and discuss the scenario with students and ask them to answer the questions. (Page 53) Task B, 'Keeping data safe inside a network'. • Talk through the answers once students have completed the exercise.	• You could relate this back to the original definition of 'firewall' in a building. The firewall cannot prevent a fire from starting, it can only contain it to one area and prevent it from spreading. • Students are probably familiar with network access permissions if they use devices connected to the school network. For example, they cannot access the records of other students, and they cannot access certain drives that teachers can access. You could demonstrate this by projecting your computer screen and showing students other drives and folders that you have access to, but they cannot even see. • Take care before you demonstrate to make sure you are not displaying any sensitive information, for example the names of files or folders that might relate to individual students.

4	• Use Build 4 (Page 71), which asks students to add questions to their troubleshooters about unauthorised access.	• You could ask the class to make some suggestions about questions to add, and compile a list on the board, for example:
		  • Can you access the website?
		  • Can you access the file?

5	• Reflection: Ask students to complete (Page 53) 'Reflection'. Students are asked to reflect on whether they think it is better to be more or less strict with security measures, giving reasons for their answers.	• It might be interesting to share some of these reflections, if students are willing to do so.
	Homework (Optional) (Page 54) • This asks students to draw a cartoon to illustrate how a firewall works .	

Build on this: Students often have passionate opinions about which websites, games etc. should be blocked or allowed on the school network. Divide the class into two sides and hold a debate.

Chapter 4:	Lesson 5 Network connections and the IoT
Student's Book page 72	
Learning Objective/s	• Describe the advantages and disadvantages of wired and wireless networks, including performance and security • List the benefits and risks of the Internet of Things (IoT)
New keywords	**Ethernet cable** – a type of network cable used to connect computers and other network devices such as routers together **Internet of Things (IoT)** – a network of interconnected devices, appliances and physical objects, typically containing sensors, which communicate across the internet **Network** – a collection of computers that are connected together to exchange data **Router** – a networking device that connects two or more networks together **Wired network** – computers connected together using physical cables such as ethernet cables **Wireless network** – computers that are connected together and can communicate via radio signals
Key pedagogy	Real-world contexts

Activity	Instruction	Guidance
1	• Ask: 'What is the difference between a wired and a wireless network?' • Ask students to complete (Page 55) Task A, 'Advantages and disadvantages of wired and wireless networks'. Students are asked to fill in the table to compare wired and wireless networks in terms of performance, security and any other considerations. • Discuss students' answers as a class once they have had time to complete the exercise.	• Students can use the material in the Student's book (Pages 72 and 73) as a starting point for this exercise, but they may also have their own experiences to draw on. • A common misconception (even among adults) is that "the Wi-Fi" is the same as "the internet". Buildings in populated areas are usually connected to the internet via a cable in the street, and then internal cables connect to a wireless router, which broadcasts the wireless network. This is not the case for 3G/4G/5G mobile networks where the signal is broadcast from a powerful antenna, or for remote areas, which may have a broadband connection via a satellite.

2	• Define the Internet of Things (Page 74) and ask students to think of examples of devices.
	• Divide students into groups for Investigate 2 (Page 74) and provide each group with a specific IoT device to research.
	• Ask students to produce a two-slide presentation (one per group), detailing the benefits their device brings, and the risks associated with the device being connected to the internet.
	• Each group should then pair up with another group and present their work.
		• Examples of IoT devices include:
		  • Smart speaker
		  • Doorbell camera
		  • Smart thermostat
		  • Smart TV
		  • Fitness tracker/smart watch
		  • Smart locks/security systems
		  • Internet connected car navigation system
		  • Healthcare devices, e.g. blood sugar monitor
		  • Public hire bikes/e-scooters

3	• Use Build 5 (Page 74). Students can now complete their troubleshooter flowcharts, ready for these to be tested during the next lesson.	• Some questions students could add regarding network connection might include:
		  • Can you connect to the device?
		  • Does the device have a password?
		  • Does your network have a password?

Activity	Instruction	Guidance
4	• Reflection: Ask students to complete (📖✏ Page 56) 'Reflection'. Students are asked about their experiences of IoT devices and their personal opinions about their security and effectiveness.	• Not all students will have IoT connected devices in their home, for a variety of reasons including cost, personal beliefs, security concerns, comfort level with technology. This is likely to be a reflection of their parents' feelings as well as their own. • Care should be taken not to present any particular viewpoint about IoT devices as being the correct one.
	Homework (Optional) (📖✏ Page 56) • Ask students to research and present a case study concerning a security issue with an IoT device. What happened?	• It is important to consider that the IoT is largely unregulated, and sometimes device manufacturers do not get things right. • Students might find suitable case studies on news websites appropriate to your country or local area, or technical news sites such as Wired.
Build on this: Should toys be connected to the internet? Discuss the risks and benefits of internet connected toys.		

Chapter 4:	Lesson 6 Troubleshooter		
Student's Book page 75			
Learning Objective/s	• Follow conditional statements in flowcharts		
You will need	• 'Troubleshooting tool test scenarios' worksheet	**Key pedagogy**	Showcase

Activity	Instruction	Guidance
1	• Tell students that in this lesson they will work with a partner to test the troubleshooting tool they have created.	• This is likely to be a busy lesson, so it is important to set the scene and provide some rough timings so that students know what to expect.

Activity	Instruction	Guidance

2	• Students may need some time to add the finishing touches to their troubleshooting tools.	• This part is optional and depends on how much time the students have had during previous lessons to work on their troubleshooting tools.
		• You could note the titles of the previous lessons on the board to prompt students to consider whether they have covered all areas in their tools:
		  • Memory
		  • The operating system
		  • Utility programs
		  • Unauthorised access
		  • Network connections and the IoT

3	• Students will use Role play 2 to test the troubleshooting tool (Page 75). Distribute 'Troubleshooting tool test scenarios' worksheet, to students. This has five pre-written computer problem scenarios for the students to solve using their troubleshooter tools. • In pairs, students will take it in turns to play the role of a computer user and a repair technician and act out the scenario. • The student playing the technician follows their troubleshooter, asking questions as they go. • Each pair records the advice they were given to solve the problem, and evaluates the quality of the response in (Page 57) Task A, 'Test the troubleshooter'.	• Not all of the questions that may be in any student's troubleshooter will be covered with answers in the role-play scenarios provided. • The student who is playing the role of the computer user should draw on their own knowledge to provide answers that are logically consistent with the scenario. • You may wish to ask each pair of students to start with a different scenario, and then work through as many scenarios as they can. This means that all scenarios will be covered by someone when you come to discuss the results at the end.

Activity	Instruction	Guidance
4	• Once the task has been completed (or there is no more time available), bring the class back together to discuss the five different scenarios and the advice that was given.	• There might be a nice opportunity here for students to openly praise other students' work by calling out specific things they thought another student did particularly well.
5	• Reflection: Ask students to complete (Page 58) 'Reflection'. Students are asked to reflect on how well the troubleshooter project went, and what they would do differently if they started the project again from scratch.	
	Homework (Optional) (Page 58) • This asks students to create a troubleshooter for a specific IoT device of their choice.	• Students may wish to use the device they researched in a group in the previous lesson as a basis for the homework task.

Build on this: Students could write their own scenarios and test them on their partner's troubleshooting tool. Can they find a scenario that the troubleshooter does not cover?

Chapter 5 Create with code 2

Project: Animal quiz

Chapter overview

Students will learn how to use pseudocode to plan their programs, and recap selection and logical operators to create a simple chatbot. They will learn how to draw truth tables for each logical operator and then bring together all of these skills in Python to create a multiple-choice animal quiz.

Chapter 5: Lesson summary		
Lesson	**Learning Objectives**	**Summary**
1	• Describe and use different data types in text-based programs, including integer, float and Boolean • Develop a text-based program that uses integer, float, string and Boolean data	Students will recap the different data types, including Boolean values. They will practise converting data between different types and write a simple chatbot program that allows different types of data to be input and output.
2	• Follow and describe algorithms that are presented as pseudocode • Explain the purpose of pseudocode, and its characteristics	Students will learn about pseudocode as a method for planning a program and compare it to a flowchart. They will then write some pseudocode for another student to translate to Python code and reflect on this process.
3	• Follow conditional statements in flowcharts and pseudocode • Develop a text-based program that uses conditional statements	Students will recap selection (if/elif/else), which they have encountered before in flowcharts and Scratch. They will then add selection statements to their chatbot program in Python.
4	• Construct truth tables for logic gates AND, OR and NOT	Students will create circuits involving AND, OR and NOT gates in a logic gate simulator, and use them to draw the truth tables. They will then use the same logical operators in a Python program.
5	• Design algorithms that include AND, OR and NOT logic	Students will research material for their quiz and plan a question in pseudocode. They will then modify the starter code given to create a quiz game in Python.
6	• Develop a text-based program that uses AND, OR and NOT logic	Students will finish their quiz program and then group up with other students to play the quizzes. They will give and receive feedback on their work.

End of chapter project	Animal quiz
Example ideas that learners could design:	• A multiple-choice quiz about animals

Why we are learning this	
Useful skills covered in this chapter	• The importance of giving precise instructions • Planning their work • How devices in everyday situations rely on logic, e.g. a car park exit barrier or a vending machine • Giving good feedback to others
Links to future careers	• Retail – calculating discounts
Links to other subjects	• Biology/Geography – animal facts • Mathematics – real numbers, percentage decrease, comparison operators, Venn diagrams

Chapter 5 – Create with code 2

Project: Animal quiz

Name	

	Working towards (1)	Meets expected (2)	Exceeding expected (3)	Score
Planning	• Shows evidence of using pseudocode to plan a simple program that allows different types of data to be input and output	• Plans a quiz question in pseudocode that includes AND, OR and NOT logic	• Uses short, clear and precise pseudocode with clear meaning to plan their quiz questions	
Use of logical operators	• Translates pseudocode into a simple Python program that allows different types of data to be input and output	• Uses the logical operators AND, OR and NOT in a simple Python program	• Constructs truth tables for logic gates AND, OR and NOT and uses them to test their program	
Multiple choice quiz program	• Modifies given code to create a simple multiple-choice quiz game in Python with multiple-choice answers that uses AND, OR and NOT logic to check whether the answer is correct	• Develops a text-based quiz program with at least four questions with multiple-choice answers that uses AND, OR and NOT logic to check whether the answer is correct and awards points to the user	• Includes some multiple-choice questions where the quiz player must select an option from a fixed list (e.g. A, B or C) and/or some validation to check whether the option typed in was valid	

Teacher feedback	
Learner response	

Chapter 5:	Lesson 1 Data types		
Student's Book page 78			
Learning Objective/s	• Describe and use different data types in text-based programs, including integer, float and Boolean • Develop a text-based program that uses integer, float, string and Boolean data		
New keywords	**Boolean** – a type of data that can have one of two possible values – true or false **Cast** – change the data type of a piece of data		
You will need	• Devices capable of running Python	**Key pedagogy**	PRIMM

Activity	Instruction	Guidance
1	• Recap data types – integer, real and string. • Use Discuss 1 (📖 Page 79). Students should look at the examples of data and decide whether each piece of data would be stored as: integer, real number or string. Give them some time to discuss this in pairs. • Discuss the results with the class. • Remind students of Boolean logic from Stage 7 and explain that Python interprets data based on how it is presented in the program (📖 Pages 78 and 79).	• Students will have covered data types in Stage 7 so a lot of this lesson will be a recap. However, they have previously used EduBlocks to support them with writing Python code, and this time they will be expected to write the code themselves into a blank code editor. • One of the examples given in the image is a telephone number. Students will often think that a telephone number should be stored as an integer because it includes numbers, but this is not correct as, while the digits are numbers, the data itself is not being used as a number, i.e. you wouldn't use a telephone number to perform a calculation. This specific example also uses a plus symbol to add the country code, which could not be represented with an integer data type. Telephone numbers should always be stored as strings.

Activity	Instruction	Guidance

2	• Demonstrate using the input function as a recap using the example (Page 79). All data is input as a string, and then changed to the data type you need.
	• Use Investigate 1 (Page 79). Ask students to predict what the output of this code will be if the user types in 2 as the input.
		• The output from the code would be:
		  • How many would you like to buy? 2
		  • The total price is 2222222222
		  • Students are likely to predict that the program will tell them the output will be 20, which is the logical intention of the program. The result is different because the input function returns data as a string. If the user inputs 2, the `people` variable contains "2" (string) and not 2 (integer).
		`10 * "2" = "2222222222"`
		`10 * 2 = 20`
		• To fix this problem, the `people` variable needs to be cast to an integer so that it can be used in a calculation.

3	• Demonstrate how to cast the value of how_many to an integer before using it in a calculation (Page 80). • Ask students to complete (Page 60) Task A, 'What type of data?'. Ask students to look at the pre-written code and fill in the table to determine the value and the data type of each variable. • Provide the answers for students to peer mark the task.	• Students have used casting in the fantasy name generator project, but this lesson covers what is happening and why in more detail. • It is possible to cast the data at the time it is input, like this: how_many = int(input("How many would you like to buy? ")) • This requires a conceptual understanding that the input statement can be considered to be equivalent to a string: input("How many would you like to buy? ") • This is because input() is a function that requests data from the user, and then returns that data as a string.

Activity	Instruction	Guidance
4	• Explain to students that they will practise input, output, concatenation and casting by creating a very simple chatbot program. They have encountered all of these skills before, but this time they will be writing Python code rather than using EduBlocks. • Ask students to complete (Page 80) Task B, 'Chatbot questions'. Ask students to plan some questions for their chatbot to ask. • Use Build 1 (Page 79). Students use these questions to create a simple chatbot program that asks questions and uses the input they typed in the reply.	• The aim of this task is to get students used to moving from a block-based language to writing fully in text in Python. Most of this lesson is a recap of material from other stages which will then be built on during the following lessons. • It is likely that students will notice the limitations of their chatbot program, especially if they have ever used a large language model before. • You could demonstrate a chatbot program that asks a question. Example code can be found in the Student's Book (Page 80).
5	• Reflection: Ask students to complete (Page 61) 'Reflection'. Students reflect on how well they remembered the recap parts of this lesson, and what they were most proud of.	• This encourages students to think about the parts of this lesson that had been covered before. How well had they remembered what they had previously learned?
	Homework (Optional) (Page 61) • Recap quiz.	• Students complete a short quiz to recap the lesson and consolidate their knowledge.

Build on this: Ask students to add a question where the answer is an integer. The chatbot should perform some kind of calculation with the integer and then use it in the reply. To do this, they will need to cast the input to an integer, and then cast it back to a string to concatenate in the response. For example:

```
age = input("How old are you? ")
age = int(age)
my_age = age * 3
print("Wow. I am " + str(my_age) + " which is 3 times older than
you!")
```

Chapter 5:	Lesson 2 Pseudocode		
Student's Book page 81			
Learning Objective/s	• Follow and describe algorithms that are presented as pseudocode • Explain the purpose of pseudocode, and its characteristics		
New keywords	**Program logic** – how the design of the program is implemented **Pseudocode** – a method of planning a program using statements that have a clear and precise meaning, but are not written in any particular programming language **Syntax** – the structure of a statement in a programming language		
You will need	• Devices capable of running Python	**Key pedagogy**	Modelling key skills

Activity	Instruction	Guidance
1	• As an introduction, talk through the material (Page 81). What is pseudocode?	• The key point here is that pseudocode is used for planning a program.
2	• Present an example of a flowchart and pseudocode for the same algorithm side by side (Page 81). • Use Discuss 2 (Page 81) • Ask students to complete (Page 62) Complete Task A, 'Similarities and differences'.	• Students will have ample experience of using flowcharts by this point in their studies, so this lesson begins by comparing and contrasting flowcharts and pseudocode as tools for planning a program. • • Pseudocode is much more likely to be used by people working as programmers because, with complex code, flowcharts can quickly become unwieldy.

Activity	Instruction	Guidance

3	• Ask students to complete (Page 61) Task B, 'Flowchart to pseudocode'. Students are given a flowchart and are asked to translate it to pseudocode
	• Use Discuss 3 (Page 82). Allow students to compare their answers with a partner – are they the same?
		• This task aims to highlight the following:
		  • A program plan must be broken into clear, short statements.
		  • The specific words used do not matter as long as the meaning is clear.
		  • The flow of the program (i.e. the order in which the statements are executed) needs to be clear.
		• This flowchart only includes statements in a sequence, but conditional statements in pseudocode will be covered in the next lesson.
		• When students compare their pseudocode with a partner, they may find differences. However, both versions can be correct as long as the meaning is clear.
		• Another benefit of comparing flowcharts with pseudocode is that if students approach the task of writing pseudocode with a blank canvas, the statements they write are often vague or far too broad (e.g. "do the calculation" or "square all the numbers"). Encourage them to think of each pseudocode statement as being similar in scope to a single box on a flowchart. You could challenge pseudocode statements with a wide scope by asking: 'Which type of box would you use to represent this line on a flowchart?'

Activity	Instruction	Guidance
4	• Ask 'Why use pseudocode when you could use Python code?' and introduce the definition of 'syntax' (Page 82). • Ask students which of the statements in the Student's Book is/are valid pseudocode (Page 82). (They all are.) • Ask students to complete (Page 62) Task C, 'Pseudocode guidelines'. Help students to write some pseudocode guidelines. You may wish to crowd-source the guidelines as a class and then have students note down the results, or you could ask students to come up with their own guidelines individually. Some example guidelines are given in the answers.	• A common question from students is why can't they just write down their plan in Python? They can! Pseudocode is used to plan the logic of a program without having to worry about knowing the exact syntax of the programming language. If students start off writing their pseudocode as Python and then realise they don't know how to do something, they can then continue to write the pseudocode with a clearly worded approximation and find out the Python syntax later. • Some students want there to be a well-defined 'syntax' or rules for pseudocode. This defeats the point of using pseudocode as it is intended to be flexible and avoid the constraint of remembering a language syntax. At later levels of study, some examination boards use a consistent format for the pseudocode in their materials so that students know what to expect in the exam paper. However, they accept answers in *any* pseudocode format provided the meaning is clear.
5	• Use Design 1 (Page 83). Students write some pseudocode for a shopping helper program. • Use Build 2 (Page 83). Students swap with a partner and use *their* pseudocode to implement the program in Python.	• Another reason you may write pseudocode as a programmer is to provide a plan for someone else to implement. Software engineers often work in pairs or teams, and so plans need to be understandable by everyone on the team, not just the person who wrote them. Students may feel uncomfortable about someone else using their pseudocode to implement a program, but it is an opportunity to experience both sides – being the author and the implementer of pseudocode. Hopefully, this exercise will provide a literal demonstration of why you should aim to write clear pseudocode!
6	• Reflection: Ask students to complete (Page 63) 'Reflection'. What is important and what is not?	• Students will often have strong opinions on whether they do or don't like using pseudocode. It is fine if, at this stage, they don't quite appreciate the benefits.

Activity	Instruction	Guidance
	Homework (Optional) (Page 63) • Students look at three examples of pseudocode and evaluate each of them by considering whether it is clear, precise and easy to understand.	• Each of the pseudocode examples given has a flaw, although some of them are easier to spot than others. Answers are provided.

Build on this: Students could give each other feedback on the pseudocode they used to implement the shopping helper program, or peer assess each other against the guidelines produced in (Page 61) Task B.

Chapter 5: Lesson 3 IF, ELIF and ELSE

Student's Book page 84

Learning Objective/s	• Follow conditional statements in flowcharts and pseudocode • Develop a text-based program that uses conditional statements		
New keywords	**Condition** – a test in a program that evaluates to a Boolean value – either true or false **Comparison operator** – an operator that allows you to compare two values, for example to check whether they are equal, or one is greater than the other **Flow of control** – the order in which the statements in a program are executed **Selection** – when a program executes different code based on a condition		
You will need	• Devices capable of running Python • Mini whiteboards (optional)	**Key pedagogy**	Investigate

Activity	Instruction	Guidance

1	• Remind students where they have seen selection statements before. There are examples in the Student's Book | (Page 84) or you may prefer to open Scratch and live-code a selection statement as a demonstration.
	• Ask students to complete (Page 64) Task A, 'Write some conditions'. As a recap, ask students to write some conditions in words. Each condition should be possible to answer with true or false.
	• After students have had a chance to complete the exercise, invite individual students to offer an example.
		• The conditions in the images are:
		  • Touching the colour red?
		  • Is it hot today?
		• Conditions can be phrased as either a statement or a question, but take care to ensure that the answer is ONLY either true or false.
		• These are examples of valid conditions:
		  • My name is Laura.
		  • Do you like rice?
		  • I am 12 years old.
		• These are *not* valid conditions:
		  • What colour is it touching?
		  • What is the weather like?
		  • How old are you?

Activity	Instruction	Guidance
2	• Introduce comparison operators in Python (Page 84). • Ask students to complete (Page 64) Task B, 'Evaluate the conditions'. This presents students with some conditions and asks whether they are true or false. • Go over the answers to the workbook task with the class.	• Students will be familiar with the comparison operators already from previous study as well as Mathematics lessons. However, their specific format in Python is important and may be new. In particular, it is worth pointing out the following: != is 'not equal to'. Students may be used to <> notation. It is common in multiple programming languages for the exclamation mark to mean 'not'. == is 'equal to'. There is often confusion between the double equals conditional operator and the single equals used for variable assignment. A good way of remembering the difference is to read == as a question: "Is it equal to?" and = as an assertion: "It IS equal to".

Activity	Instruction	Guidance
3	• Demonstrate the syntax of an if/else statement in Python (Page 85). • Use Build 3 (Page 85). Students add three more questions to their chatbots. Each question should use a condition to give a different response depending on what the user typed.	• For a practical demonstration of the concept of 'flow of control' you could try this activity: 1. Print out each line from the example program in the Student's Book (Page 85) on a separate piece of paper (or you could use mini whiteboards if you have them). 2. Ask five students to come to the front and, facing the class, each hold up a statement, lined up in the correct order. 3. Have another piece of paper labelled "flow of control", which starts off with the student holding the first statement. 4. Students 'execute' their statements in turn, asking for input from the class where required. The flow of control paper should be passed to the student who should execute their statement next. • If the input from the class is water, statements 1, 2, 3 should execute and if it is anything else statements 1, 4, 5 should execute. Use this to discuss how the selection statement causes the program to flow differently depending on whether the condition was true or false.
4	• While writing their questions, students may have encountered a situation where they wanted to have more than one specific response. Introduce elif syntax in Python (Page 86). • Introduce Build 4 (Page 86), where students add a question with an elif to their program (they could also modify an existing question if they prefer).	• The elif statement shown orders the conditions in a specific way: >=40, then >18, then the catch-all else condition. The conditions are evaluated *in order*, and, as soon as one condition matches, ALL the rest of this selection statement is skipped. Therefore, the order of the conditions is important. The condition >=40 has to come before >18 because it is more specific. This point is covered explicitly in the homework task for this lesson.

Activity	Instruction	Guidance
5	• Reflection: Ask students to complete (Page 64) 'Reflection'. This asks students whether there was anything they didn't know how to do and how they would find out how to do something they are stuck on.	• Teaching students at an early stage to have some strategies for being self-sufficient programmers will help you to provide the best targeted support for all learners in the class. For example, students could ask a friend, look back at their notes or their book, search online or even just try something out and investigate what happens, before asking you for help.
	Homework (Optional) (Page 65) • Students trace the Python code and write down the output generated by the inputs specified.	• This task uses the 'investigate' part of the PRIMM pedagogical approach to programming. Can students look at a piece of code and trace it to find out what it does?

Build on this: Python has another statement that can be used for selection, called **match** (in other programming languages this is often called 'switch'). The variable to check is written after **match** and the possible values of the variable are written as each **case**, with the catch-all equivalent of the else statement written as **case _**, at the end.

```
colour = input("What is your favourite colour? ")
match colour:
    case "green":
        print("Like the grass!")
    case "blue":
        print("Like the sky")
    case _:
        print("That's a nice colour")
```

You could display this code and ask students to copy it and then add more outputs for other different colours. They could also test their program works as they intended. What if someone types "Green"?

Chapter 5:	Lesson 4 AND, OR, NOT
Student's Book page 87	
Learning Objective/s	• Construct truth tables for logic gates AND, OR and NOT
New keywords	**Logic gate** – a circuit inside a computer allows Boolean logic to be applied to one or more inputs **Truth table** – a diagram showing all possible combinations of inputs and outputs from a Boolean expression
You will need	• Devices capable of running Python • A logic gate simulator program (see guidance)
Key pedagogy	Real-world concepts

Activity	Instruction	Guidance

1	• Ask students whether they can recall logic gate symbols for AND, OR and NOT. You could perhaps ask for a volunteer to draw each gate on the whiteboard (Page 87).
	• Recap the logic of each gate, and that inputs/outputs can be either true or false, which is usually noted as 1 or 0.
	• Ask students to complete (Page 66) Task A, 'Recap logic gates'. Students label the logic gate diagrams in their Workbooks.
		• Students have encountered logic gates already in Stage 7, so this is a chance to recap the three different gates they are familiar with.
		• If it aids recall, you can think of 0 as "nothing" or "no", which links to false.
		  • NOT gate – the output is the opposite of the input
		  • AND gate – both inputs must be true (1) for the output to be true (1)
		  • OR gate – one or both inputs must be true (1) for the output to be true (1)
		• Highlight that the NOT gate has only one input, but the AND and OR gates both have two inputs (but one output).

2	• Introduce truth tables as a diagram showing all possible combinations of inputs and outputs. • Demonstrate filling in the truth table for the NOT gate using a logic gate simulator (Pages 87 and 88). • Ask students to complete (Page 66) Task B, 'Complete the truth tables'. Students create the logic gates AND and OR in a logic gate simulator and then complete the truth tables.	• You can find logic gate simulators online that are free to use and don't require an account. We used the one at (Website C5L4R1). • Alternatively, there is a scratch project from Stage 7, Chapter 3 at (Website C5L4R2), which allows students to test the logic using pre-created gates.

Activity	Instruction	Guidance
3	• Use Discuss 4 (📖 Page 88). Ask students to discuss in groups and predict which logic gate would be represented by the truth table for the real-world contexts. • Draw each truth table on the board together with the class and check the answers.	• Answers, including truth tables, for the discussion task are provided. • Some students may notice that the answer to each scenario is actually included in the description, if you look carefully for the logical operator "and", "or" or "not". • With the fridge scenario, which is an OR gate, highlight that the alarm will still go off if both inputs are true, i.e. if the door is left open and the temperature goes above 8°C. The output of an OR gate will still be true if both inputs are true. This is different from how we use the word 'or' in everyday language, e.g. "would you like tea or coffee" – the answer "both" is not usually acceptable! There is a separate gate that covers this logic called an XOR (exclusive or) gate, but this is not a gate that students need to know at this stage. • Fun fact: there is only one word for "or" in English, but in some languages there are different words for OR and XOR!

Activity	Instruction	Guidance
4	• Explain the syntax for the operators in Python (Page 89). • Work through Build 5 where students use their knowledge of selection statements in Python from the previous lesson, combined with the logical operators, to write a program asking for a password. • Progress to Build 6 where students build on their previous programs by combining two logical operators. • The answers to both build tasks are provided in the answer document. After students have had a chance to write their own programs, you could either provide the sample solutions for students to look at, or go over them with the class.	• The syntax in Python is extremely straightforward as it is simply the logical operator words in lowercase. • Conditions in Python can use brackets to make meaning explicit, in the same way as they are used in mathematical equations. Adding brackets around the whole condition allows you to split it across multiple lines, which is extremely useful for long conditions. • Some students may realise that adding each user account as an additional condition in this way is not how this would be implemented in a real application. You could ask them extension questions such as: 'How do you think a real programmer might do this?' (Answer: The username/password combinations would probably be stored in a database.) and: 'What problems can you see with this approach?' (Answer: Each time a new user is added, the code would have to be modified; if someone had the code they would know everyone's password etc.)
5	• Reflection: Ask students to complete (Page 67) 'Reflection'. Students reflect on Boolean notation and any difficulties they had during the lesson.	• Sometimes students may find a particular notation more difficult to remember. It is fine to use any notation (true/on/1 and false/off/0) if that makes the concepts easier to understand.

Activity	Instruction	Guidance
	Homework (Optional) (Page 67) • Create a truth table for a more complicated circuit that uses two logic gates joined together.	• This is an extension task because it looks at a circuit that combines two logic gates. Students will be expected to be able to combine logic gates by themselves at later stages, but this is a very straightforward example. • Be careful when marking question 4 – the answer "when both inputs are false" is *incorrect*. • This circuit can actually be drawn using a logic gate, which students do not need to know at this stage. The gate is called a NAND ("not and") gate, because the truth table is the reverse of an AND gate.

Build on this: Ask students to think of their own real-world context and to draw a truth table for that context, in a similar way to the examples in Discuss 4 (Page 88).

It is possible that some students may come up with more complicated scenarios than can be covered using a single logic gate – can they work out how to combine logic gates to achieve the scenario they want?

Chapter 5: Lesson 5 Design and build a quiz program

Student's Book page 90			
Learning Objective/s	• Design algorithms that include AND, OR and NOT logic		
New keywords	**Indentation** – positioning code further in from the left **Logical operator** – AND, OR or NOT **Sequence** – statements in code that execute in the order in which they are written		
You will need	• Devices capable of running Python • Starter code for quiz	**Key pedagogy**	Investigate

Activity	Instruction	Guidance

1	• Use Discuss 5 (Page 88): Discuss 5. Ask students to discuss with a partner what the output from the pseudocode would be if the input is 5.
	• Bring the class back together and ask for the answer. There are two possible answers, and it is not possible to distinguish between them with the code in its current form (see guidance).
	• Demonstrate how to write the pseudocode so that the meaning is explicit (see guidance).
		• Talk through the two possibilities as described below. The key point to stress is that it *must* be clear in pseudocode which statements are only executed when the condition is met, and which are not. You may wish to write the code on the board and then dry run it line by line to demonstrate.
		• It is unclear whether both output statements are part of the condition, or only the first one. The code could have one of two interpretations.
		1. Only the first output statement is part of the condition, so the output for input 5 would be "Well done".
		`INPUT score`
		`IF score > 10`
		`OUTPUT "Excellent score"`
		`OUTPUT "Well done"`
		2. Both statements are part of the condition, so the output for input 5 would be nothing.
		`INPUT score`
		`IF score > 10`
		`OUTPUT "Excellent score"`
		`OUTPUT "Well done"`
		• Students *must* be clear in their pseudocode (and in Python), either by using indentation, or by using ENDIF statements (see (Pages 90 and 91).

<table>
<tr><th>Activity</th><th>Instruction</th><th>Guidance</th></tr>
<tr><td>2</td><td><ul><li>Introduce the focus of the next two lessons – students will create a multiple-choice animal quiz, but each question requires a logical operator (AND, OR, NOT) to check the answer (Page 91).</li><li>Ask students to complete (Page 68) Task A, ‘Finish the pseudocode’ to check the answer for the quiz question provided.</li><li>After students have had a chance to complete the task, go over the answer in pseudocode with the class to check that students have understood.</li></ul></td><td><ul><li>An answer to the Workbook task has been provided in the answer section.</li></ul></td></tr>
<tr><td>3</td><td><ul><li>Ask students to complete (Page 69) Task B, ‘Plan your quiz questions’. Students plan the questions for their quiz.</li><li>There are some examples provided for how to write questions that use each operator (Page 92).</li></ul></td><td><ul><li>If necessary, you can provide some guidance if students' pseudocode plans have any obvious errors, for example:<ul><li>lacking or incorrect indentation</li><li>did not use sensible variable names.</li></ul></li><li>We have suggested eight questions, but you may want to set a time limit of approximately 10–15 minutes for this section of the lesson, so that students have enough time for the programming task.</li></ul></td></tr>
<tr><td>4</td><td><ul><li>Provide students with the starter code for a quiz, which they can modify to add their own questions and logic.</li><li>Ask students to go through each point in the project brief and use it to develop their quiz programs (Page 92).</li></ul></td><td><ul><li>The starter code can be found at starter_code.py</li><li>The starter code includes a variable for a score. This is discussed in the next lesson, but some students may add code to increase the score as they go.</li></ul> To implement a NOT question in Python, the Boolean operator must come first to negate the result of the condition: <pre><code>if not (answer == "A"):
    print("Wrong answer")</code></pre> You may prefer to suggest that students use the comparison operator != (not equal to) to achieve the same result in a more intuitive way: <pre><code>if answer != "A":
    print("Wrong answer")</code></pre></td></tr>
</table>

Activity	Instruction	Guidance
5	• Reflection: Ask students to complete (📖✏ Page 69) 'Reflection'. Students are asked to reflect on the use of pseudocode to plan their questions before writing them in Python code.	• This is the second time that students will have used pseudocode to help to plan a program, so the reflection task aims to check in and see whether they are finding the planning stage helpful.
	Homework (Optional) (📖✏ Page 70) • Students need to annotate the code.	• This task encourages students to 'read before they write' by showing them some example code. They must then label the different parts of the code. You can use this task to assess how well students have understood both the terminology that has been covered, and their ability to read code and work out what it does.

Build on this: Students could vary their questions so that they also include some multiple-choice questions where the quiz player must select an option from a fixed list (e.g. A, B or C). They could add some validation to check whether the option typed in was valid.

Here is some Python code that prompts the player to type A, B or C, and then if their **choice** was not in the list **["A", "B", "C"]** they will be prompted to re-enter a choice until it is one of the valid options.

```
choice = input("Type A, B or C ")
while choice not in ["A", "B", "C"]:
    choice = input("Type A, B or C ")
```

This code uses iteration and lists which are not required knowledge at Stage 8, but students can use it as part of their program and even modify it (as per the PRIMM pedagogical strategy), for example they may wish to include an option "D" in the list.

Chapter 5: Lesson 6 Showcase your quiz

Student's Book page 93

Learning Objective/s	• Develop a text-based program that uses AND, OR and NOT logic		
You will need	• One copy of the 'Quiz feedback worksheet' per student • Scissors (optional)	**Key pedagogy**	Showcase

Activity	Instruction	Guidance
1	• Mention the **score** variable that is included in the starter code students were given (Page 93). • Use Build 7 (Page 93). Students finish implementing the programs for their quizzes, including adding points to the score if a question is answered correctly.	• As an extension for more advanced students, you could ask them to see whether they could award half marks for an answer that was partially correct. For example, if a question was an "AND" question and they got one of the parts correct but not the other, can they add this as a test in their program logic and award half a mark?
2	• Ask students to showcase their quizzes (Page 93). Organise the students into groups of three or four. Each group member plays all quizzes in the group and can also look at the code the other students have written. • Ask students to record their scores and notes in (Page 71) Task A, 'Quiz time'.	• Keep the group sizes quite small otherwise students will have too much to do. If your students needed a lot of time during this lesson to finish their programming, you may want to do the feedback in pairs. You may also want to strategically group students so that they are similar in ability, so their quiz programs will be similar in quality. • As usual, when students are using computers logged on as another student to test work, remind them to behave appropriately. Alternatively, if you have a shared work area available, you can ask students to upload a (named) copy of their quiz code to the shared area and then students remain in their own seats to test the quizzes from their group.
3	• Each student writes feedback on the 'Quiz feedback' worksheet for the students whose quizzes they tested. • Cut out the individual feedback slips and provide them to the correct students.	• You could create a classroom display by printing out some of the students' work and then displaying their feedback slips around the program code and/or output.
4	• Reflection: Ask students to complete (Page 72) 'Reflection'. Students are asked to consider how programming their quizzes could have been made easier and to say what they admired in another student's quiz. They will also reflect on how well they met the project brief.	• Students' feelings may vary about the planning process involved in this chapter. Some will enjoy the structure of planning, and some will feel frustrated by it.

Activity	Instruction	Guidance
	Homework (Optional) (Page 72) • Students are asked to write an expression using a logical operator for the shaded area in the Venn diagram.	• Students will have encountered Venn diagrams before in Mathematics lessons. It may be interesting for them to realise that despite sounding more inclusive, AND is more restrictive than OR because both criteria have to be met. This is very clearly demonstrated with a Venn diagram.

Build on this: Continue to encourage students to use pseudocode to plan their programs. It is often a required step for coursework or controlled assessments for future qualifications and so it is a good idea to get into good habits early on.

Chapter 6 Connect the world

Project: Virtual pen pal

Chapter overview

In years gone by, people enjoyed writing letters to each other and sending them in the post. Now lots of people communicate via the internet, using technology such as instant messages and emails. How can a computer get a message to your pen pal on the other side of the world?

Chapter 6: Lesson summary		
Lesson	**Learning Objectives**	**Summary**
1	• Convert positive whole numbers between binary and denary	Students will learn how numbers are represented in binary and convert between binary and denary numbers.
2	• Compare copper cables to fibre optic cables as a means of data transmission • Describe types of network, including PAN, LAN and WAN	Students will look at two different types of cable used to carry data across a network. They will then consider the different sizes of network, ranging from their personal network to the internet.
3	• Describe how ASCII is used to represent letters and characters	Students will learn why an agreed text format called ASCII was created. They will then use ASCII to represent words and complete a practical activity to communicate a message with beads.
4	• List the benefits and limitations of online communication media	Students will start by listing the different ways in which we can communicate online and will then evaluate the benefits and limitations of each method in a range of scenarios.
5	• Explain how an echo check works • Define compression and describe why it is needed	Students will role-play an echo check, and then use a compression technique called run-length encoding to compress a message and send it again.
6	• Showcase and explain a piece of work to peers	Students add the finishing touches to their booklets during this lesson. They then showcase their work to their peers, explaining the part they are most proud of. Each student will peer evaluate another student's work and offer feedback.

End of chapter project	**Virtual pen pal**
Example ideas that learners could design:	• A booklet telling the story of how a message can be sent instantly to a friend in another country

Why we are learning this	
Useful skills covered in this chapter	• Some files can be compressed to save space • Presenting and explaining a piece of work to peers • Offering useful feedback to a peer
Links to future careers	• Marketing – different online communication methods
Links to other subjects	• Mathematics – number bases • Modern foreign languages – how characters from different languages are encoded on a computer • Science – electromagnetic interference, materials used to make cables

Chapter 6 – Connect the world

Project: Virtual pen pal

Name	

	Working towards (1)	**Meets expected (2)**	**Exceeding expected (3)**	**Score**
How data is transmitted	• Understands why computers use binary to transmit data	• Understands the text format called ASCII and why it was created and explains how ASCII is used to represent words	• Compares the two different types of cables used to carry data across a network and considers the different sizes of network, ranging from a personal network to the internet	
Online communication	• Describes the benefits and limitations of one method of online communication	• Lists the different ways in which we can communicate online, and then evaluates the benefits and limitations of each method in a range of scenarios	• Explains how run-length encoding is used to compress data and then how data that has been run-length encoded can be uncompressed	
Present a booklet	• Produces a booklet telling the story of how a message can be sent instantly to a friend in another country that focuses on one communication method	• Produces a booklet explaining how instant messaging works, which includes information about how an echo check works and compression	• Produces a booklet explaining in detail how instant messaging works, which includes photographs of their pixel pictures alongside their run-length encoding	

Teacher feedback	
Learner response	

<table>
<tr><th colspan="5">Chapter 6: Lesson 1 – Binary and denary</th></tr>
<tr><td colspan="5">Student's Book, page 96</td></tr>
<tr><td>Learning Objective/s</td><td colspan="4">• Convert positive whole numbers between binary and denary</td></tr>
<tr><td>New keywords</td><td colspan="4">Binary – a number system that represents numbers using only two values: 0 and 1
Bit – a single binary digit, either 0 or 1
Denary – the familiar base-10 number system, which uses the digits 0–9
Place value – the value that a digit has because of its position ('place') in the number</td></tr>
<tr><td>You will need</td><td colspan="2">• Markers – washable
• Stickers (optional)
• A4 paper</td><td>Key pedagogy</td><td>Lead with concepts / Explain to others</td></tr>
</table>

<table>
<tr><th>Activity</th><th>Instruction</th><th>Guidance</th></tr>
<tr><td>1</td><td><ul>
<li>Use Discuss 1 (📖 Page 96). Ask: 'Can you count to over 1000 just using your fingers?' The answer is yes, if you count in binary .</li>
<li>Ask students to label their fingers as shown either using washable markers or with stickers. Demonstrate how to make the number 0 – a closed fist with no fingers showing, and then the number 1 – you hold out only your thumb (📖 Page 96).</li>
<li>Students look at the examples in the Student's Book and decide which number is represented by each picture.</li>
<li>Use Investigate 1 (📖 Page 97). Students work in pairs and take turns to show a number or calculate a number given by their partner (📖 Page 97).</li>
</ul></td><td><ul>
<li>Students have briefly encountered binary representation in Stage 7 Chapter 4.</li>
<li>When students are labelling their fingers, it is important that they label with the smallest place value (i.e. 1) the furthest to the right. It may be hard for right-handed students to label their own right hands, so they could pair up with a partner and label each other's hands. We have deliberately advised labelling the right hand, so that you can potentially go on to label the left hand with higher place values to extend the range of numbers available.</li>
<li>If any students do use their left hand to label, the place values should remain in the same order, so, with the palm facing you, the smallest finger should be 1 and the thumb should be 16.</li>
<li>It is important to avoid labelling the place values with the smallest place at the left as this could result in a misconception later on.</li>
<li>Picture answers<ul><li>2 + 1 = 3</li><li>16 + 2 + 1 = 19</li></ul></li>
<li>The largest possible number you can represent using two hands is 1023.</li>
</ul></td></tr>
</table>

<table>
<tr><th>Activity</th><th>Instruction</th><th>Guidance</th></tr>
<tr><td>2</td><td><ul><li>Students have previously learned about logic gates that take true or false inputs. Draw a comparison between these and binary where there are only two possible digits.</li><li>Computers store numbers as binary because they can represent two values, using a high or low voltage (Page 97).</li></ul></td><td><ul><li>Computers are made up of billions of switches which are either on (1) or Off (0)</li><li>This may be the first time students have really thought about the concept of there being other ways to write down numbers / other number bases.</li><li>Understanding that binary digits can only ever be 0 or 1 also helps avoid misconceptions in later study; for example, when adding binary numbers students will sometimes wrongly introduce a 2 into their calculations.</li></ul></td></tr>
<tr><td>3</td><td><ul><li>Use the explanation (Page 98) to demonstrate a worked example of converting a binary number to denary on the board.</li><li>Ask students to complete (Page 74) Task A, ‘Binary to denary conversion’.</li></ul></td><td><ul><li>Again, choose a number with a maximum of 5 bits to avoid making the conversion too difficult in the demonstration.</li><li>Note that we have used 5 bits in these examples to correspond with the 5 fingers the students initially used, but the binary numbers students are likely to encounter in future study will be most likely to use 8 bits. There are some 8-bit conversions in the homework task.</li><li>Converting larger numbers works in exactly the same way. Continue to label the place value headers beginning with 1 as the place value furthest to the right and then doubling each time.</li></ul>
<table>
<tr><td>128</td><td>64</td><td>32</td><td>16</td><td>8</td><td>4</td><td>2</td><td>1</td></tr>
<tr><td>1</td><td>0</td><td>1</td><td>1</td><td>0</td><td>1</td><td>0</td><td>0</td></tr>
</table>
$128 + 32 + 16 + 4 = 180_{10}$</td></tr>
</table>

<table>
<tr><th>Activity</th><th>Instruction</th><th>Guidance</th></tr>
<tr><td>4</td><td><ul><li>Use the explanation (Pages 98 and 99) to demonstrate a worked example of converting a denary number to binary on the board.</li><li>Ask students to complete (Page 74) Task B, 'Denary to binary conversion'.</li><li>Ensure that students do not use devices during this part of the lesson – number base converters are widely available on the web and you want them to practise converting the numbers manually.</li></ul></td><td><ul><li>The correct term for the base-10 number system that is widely used is denary, although students may also have heard it referred to as decimal. The term denary is used here to avoid confusion between the number base and numbers with decimal points.</li><li>If you choose to demonstrate a different number to the one shown in the Student's Book, the number you choose should be 31 or less, otherwise you will require more than 5 bits to represent it.</li><li>Here is another worked example:</li><li>21 in denary is 10101 in binary.</li></ul>
Target: 21
<table><tr><td>16</td><td>8</td><td>4</td><td>2</td><td>1</td></tr><tr><td>1</td><td></td><td></td><td></td><td></td></tr></table>
Target: 21 – 16 = 5
<table><tr><td>16</td><td>8</td><td>4</td><td>2</td><td>1</td></tr><tr><td>1</td><td>0</td><td></td><td></td><td></td></tr></table>
Target: 5
<table><tr><td>16</td><td>8</td><td>4</td><td>2</td><td>1</td></tr><tr><td>1</td><td>0</td><td>1</td><td></td><td></td></tr></table>
Target: 5 – 4 = 1
<table><tr><td>16</td><td>8</td><td>4</td><td>2</td><td>1</td></tr><tr><td>1</td><td>0</td><td>1</td><td>0</td><td></td></tr></table>
Target: 1
<table><tr><td>16</td><td>8</td><td>4</td><td>2</td><td>1</td></tr><tr><td>1</td><td>0</td><td>1</td><td>0</td><td>1</td></tr></table>
</td></tr>
</table>

Activity	Instruction	Guidance
5	• Use Discuss 2 (Page 100). Introduce the terminology for 'bit' and ask students what is the largest number you can store with 5 bits? What happens if the number you want to store is too large? • Demonstrate attempting to convert a number 32 or larger from denary to binary, adding an extra bit with a place value of 32. • Use Discuss 3 (Page 101). Ask: 'Why do you need to double the largest place value and subtract 1 to find the largest possible number you can represent?' • Discuss with the class that computers store *all* data as binary digits, not just numbers. This will be important in subsequent lessons.	• The term 'bit' is actually short for **b**inary dig**it**! • 5 bits allows for 32 different permutations. One of the permutations stores the number 0, so the *largest* number that can be stored is always one less than the number of permutations. • Depending on how confident your students are mathematically, you could explain that the place values are powers of 2. $2^0 = 1$ $2^1 = 2$ $2^2 = 4$ $2^3 = 8$ $2^4 = 16$ $2^5 = 32$ etc.
6	• Tell students that during this project they will create a paper booklet to explain the story of a message being sent to their 'virtual pen pal' on the other side of the world. • Ask students to complete Build 1 (Page 101). Students are asked to add explanations to cover part of the project brief in their booklet: converting numbers between binary and denary, and explaining why computers use binary to transmit data.	• At this stage, students do not need to worry about a front cover, layout, page numbers, etc. It is more important to ensure that they note down a clear explanation of what they have covered during the lesson. • If students are unfamiliar with the concept of a 'pen pal' in the traditional sense (a friend who you exchange letters with in the mail), you might want to spend some time exploring this concept to set the scene for the rest of the chapter, and throughout the chapter you can draw comparisons with the ways people used to communicate ~20–30 years ago compared with modern communication.
7	• Reflection: Ask students to complete (Page 75) 'Reflection'. Students are asked to reflect on prior knowledge, and why they are learning about this topic.	

Activity	Instruction	Guidance
	Homework (Optional) (Page 75) • Students practise converting larger numbers (up to 8 bits).	• In this task, if the number has subscript $_{10}$ it means that the number is in denary, and if the number has subscript 2 it means that the number is in binary. • Accept leading zeros as part of an answer, e.g. the answer 1010 is equivalent to 00001010.

Build on this: There is an alternative method for converting denary to binary, but it requires whole number division (//) and knowledge of remainders so it is not used in the Student's Book. Here is a worked example using the number 6.

1. Divide the target number by 2 and write the remainder. Repeat until the target number reaches 0.	6 // 2 = 3, remainder 0 Write a 0, and the new target number becomes 3 3 // 2 = 1, remainder 1 Write a 1, and the new target number becomes 1 1 // 2 = 0, remainder 1 Write a 1, and the new target number becomes 0
2. Write the remainders in reverse to get the binary number.	Result (written backwards): 110 = 6 in binary.

Chapter 6:	Lesson 2 How are we connected?		
Student's Book page 102			
Learning Objective/s	• Describe types of network, including PAN, LAN, and WAN • Compare copper cables to fibre optic cables as a means of data transmission		
New keywords	**Fibre optic cable** – a type of cable that uses flexible glass fibres to transmit data as pulses of light **Copper cable** – a type of cable that uses copper wire to transmit an electrical signal **Interference** – disruption or alteration to an electrical signal **Transmit** – send a signal from one place to another **WAN** – Wide Area Network, a network that connects devices over a large geographical area, for example the internet **LAN** – Local Area Network, a network that connects devices in one physical location, for example a school or an office **PAN** – Personal Area Network, a network that connects the devices within range of one person		
You will need	• Ethernet cable (optional)	**Key pedagogy**	Real-world contexts

Activity	Instruction	Guidance
1	• Ask the class: 'When you plug in a network cable, how is the data transmitted?' • Talk through the information (📖 Pages 102 and 103). There are two main types of network cable: copper cables and fibre-optic cables. • Ask students to complete (📓✏ Page 76) Task A, 'Copper and fibre optic cables'. Students are asked to label the parts of each type of cable on a diagram and note down the advantages and disadvantages of each.	Students have learned about wired vs wireless networks in Chapter 4.5. To provide an audible example, this page has a sound file of interference with a speaker caused by a mobile phone signal. (Website C6L2R1)

Activity	Instruction	Guidance
2	• Use Discuss 4 (Page 103). Ask students to discuss in pairs. • Once students have had time to discuss this, ask them to share their thoughts with the class.	• Students may realise that there is not one single cable connecting them and another person. They may realise that they connect to their home or school network either with a cable or wirelessly, and that they are actually connecting to a device such as a home router. That device is then connected to an external network such as the internet, probably via a cable that comes directly into the building. If you have a spare ethernet cable, you could pass it around for students to look at. • You could remind them of what they learned in Chapter 4.4 about firewalls – as they will see next, different networks can connect together. The internet is actually multiple networks, all connected together.
3	• Talk through types of network (Pages 103 and 104) – WAN (internet), LAN (school), PAN (e.g. your headphones). • Ask students to complete (Page 76) Task B, 'PAN, LAN or WAN?' Students are given some characteristics of networks and asked to tick whether each one is true for a PAN, LAN or WAN. • Talk through the answers with the class. You could ask students to swap and mark each other's work.	• Many devices that form part of a PAN are probably connected via Bluetooth. Students may recognise this as a short-range wireless connection option. • Some of the statements in the Workbook task apply to more than one of the sizes of network. • You could provide some specific context for your students in terms of your school's own network. For example, where does the broadband connection come into the school? Do they save files to the local network? You may be able to ask one of your network technicians to visit your lesson briefly to give some context.
4	• Explain that students will complete Build 2 in their project booklets (Page 104).	• Once again, students should focus on adding the information they need from the lesson to their booklet. There will be time during the final lesson to add finishing touches, or to complete any diagrams or illustrations.

Activity	Instruction	Guidance
5	• Reflection: Ask students to complete (Page 77) 'Reflection'. Students reflect on how they could use what they have learned outside of the lesson and think about how it relates to their own environment.	• Relating to real-world contexts helps students to remember the material from the lesson and makes it more meaningful to their own life.
	Homework (Optional) (Page 77) • Students are asked to draw a network, or networks, they are connected to, explain the type of network and the devices that are connected to it.	• Students could draw these by hand in their Workbook, or they could find appropriate images online and copy/paste them into the Workbook before annotating them if they feel less confident.

Build on this: Ask students to research another type of network called a Metropolitan Area Network, or MAN. This is a network that is larger than a LAN but smaller than a WAN, and it connects devices within a large area such as a whole city, or the whole campus of a university.

Chapter 6: Lesson 3 ASCII

Student's Book page 105

Learning Objective/s	• Describe how ASCII is used to represent letters and characters		
New keywords	**ASCII** – American Standard Code for Information Interchange, a commonly used format for encoding characters on a computer **Unicode** – a character encoding format that can encode characters in any language		
You will need	• String • Beads, tube shaped pasta or straws in two different colours • 'ASCII values for common characters' worksheet • Calculators (optional) • 'Unicode' worksheet (optional)	**Key pedagogy**	Unplugged

Activity	Instruction	Guidance

1	• Divide students into groups and explain Investigate 2 (Page 105):
	  • They need to communicate a single English word across the room.
	  • One person is assigned as the communicator, and this person may only use **one hand** to communicate the word.
	  • They will **not** know the word in advance.
	  • Students have an initial 5 minutes to work together to invent a method of communication.
	• Once the time is up, ask the students who are communicators to move to one side of the room, and secretly provide each of those students with a single word.
	• The communicator then attempts to communicate that word using the method they have pre-agreed with their group.
		• Use a different word for each group to make sure that the groups cannot benefit from each other's methods!
		• The word is deliberately specified as being an *English* word. Some students may be bilingual and when they study ASCII later they may be quick to notice that characters from languages other than English are missing from the ASCII character set. This is discussed in the 'Unicode' worksheet.
		• When students are attempting to communicate their word, they should be completely silent, and remind them not to try to communicate in any other way, for example, by using facial expressions.
		• Some students may choose to represent letters using binary numbers, thinking back to the exercise they did in the first lesson of this chapter. Others may choose different methods, for example some may know sign language, or some may try to form letter shapes with their hands. The method does not matter; the key point is that the method should have been *agreed* in advance, so that both sides know what to expect.

2	• Reflection: Ask students to complete (Page 78) 'Reflection'. Students reflect on how successful their communication was, and why. • Ask some students to share their reflections with the class.	• At this point, try to draw out the point that the students needed to have agreed a set of rules for the communication in advance, so that both the communicator and the group trying to receive the word knew what was meant by the communication.

Activity	Instruction	Guidance
3	• Introduce ASCII (Page 105). Each letter in the Latin alphabet (as well as other characters such as punctuation) has a corresponding number code that can be stored as binary. • Ask students to complete (Page 79) Task A, 'Write your name using ASCII'. Students use the 'ASCII values for common characters' worksheet to look up characters and write their names in ASCII.	• Make sure that students distinguish correctly between upper and lowercase values as they have different ASCII codes. • They should give the binary for each character as an 8-bit binary number. They may find it easier to represent the number as two lots of 4 bits, for example a lowercase 'a' is 97 which is 0110 0001 as an 8-bit binary number. ASCII originally used 7 bits, but since 8 bits is a common unit (a byte), it was later revised as 8-bit extended ASCII with more characters able to be represented. • Some binary trivia – 8 bits is called a byte, and half a byte, i.e. 4 bits, is called a nibble!

Activity	Instruction	Guidance
4	• Introduce the Build tasks and explain that students will create a piece of jewellery to represent their name in ASCII (Page 106). Provide students with the string, beads or other materials you have collected. Ask them to thread the beads to create a string of beads containing their name in ASCII. • Once the item has been created, students can swap with each other to check that the ASCII codes have been added correctly.	• For this exercise, you will need small beads in two colours, preferably one lighter and one darker colour. If you do not have beads, you could provide tube shaped pasta and markers to colour it, or cut up drinking straws in two different colours to thread onto the string. • If required, students can look up other characters not included on the 'ASCII values for common characters' worksheet. There are ASCII tables with various punctuation widely available on the web, for example (Website C6L3R1). • It is important that students consistently use 8 bits for each ASCII letter, otherwise their message will be very difficult for someone else to decode. They may also want to mark the start of the message in some way to distinguish it from the end; for example, by knotting the string twice, or by using a single bead of a different colour. • Example: G = green, Y = yellow and R = red: G = Line break, Y = 0 and R = 1 Beads are threaded in the following order (line breaks have been used here to make each character clear). G YRYRYRYY YRYYYYYR YRRRRYYR YRRYRRRR The student's name is Tayo.

Activity	Instruction	Guidance

5	• Use Build 4 (Page 106). In their project books, students should add an explanation of the following objectives:
	  • Demonstrate how to write a word using ASCII.
	  • Convert ASCII codes back to letters and other characters.
	• (Optional) Students could use a camera to take a photograph of the string of beads they created, to use in their project book.
		• If students are able to use a photograph of their own work from the lesson in the book, this might be a nice way of remembering the activity while reusing the materials. If it is feasible, you may also wish to allow students to keep the messages they have made.

	Homework (Optional) (Page 79) • Students need to work out the secret message written in ASCII.	• This task gives students the opportunity to convert back from an ASCII code to a letter.

Build on this: Unicode is a different character encoding system with many more possibilities. Unicode character values are often written in hexadecimal (base-16), for example U+1F602 is a laughing face emoji 😂, because the equivalent number in binary would be very long. Hexadecimal is a different number base system, but whereas binary has two different possible digits (0 and 1), hexadecimal has 16 possibilities – the numbers 0-9 plus the letters A-F.

There is an optional worksheet 'Unicode' with further background information about Unicode and a short task.

Chapter 6:	Lesson 4 Online communication		
Student's Book page 107			
Learning Objective/s	• List the benefits and limitations of online communication media		
New keywords	**Instant message** – a short text-based message that is delivered immediately, allowing them to be used for a real-time conversation **Live stream** – to broadcast an online event as it happens, for example, someone playing a video game, or presenting to the camera **Social media** – a website or an app that allows users to share text, image and video content with others they are connected to **Voice chat** – a method of online communication via speech, often involving joining a channel with a specified theme		
You will need	• Large sheets of paper • Marker pens or felt tips in six different colours	**Key pedagogy**	Collaboration

Activity	Instruction	Guidance
1	• Use Discuss 5 (Page 107). Start off with a class discussion. Ask students to name online communication media you could use to send a message to someone in another country. • Talk through any of the media on that may have been missed (Pages 107, 108 and 109).	• The main methods that will be discussed during this lesson will be instant messaging, social media, video calls, voice chat, online games and email. Other methods of online communication exist (see 'Build on this' section).

Activity	Instruction	Guidance

2	• Use Investigate 3 (Page 109). Divide students into six groups and give each group a different coloured marker pen. Allocate each group one of the following contexts:
	  • School
	  • Friends your age
	  • Extended family
	• Write a title on six different large pieces of paper, one for each type of communication method:
	  • Instant messaging
	  • Social media
	  • Video call
	  • Voice chat
	  • Online game
	  • Email
	• Each group receives one piece of paper and is given 5 minutes to write down benefits and limitations of each communication method that are relevant to their context. They should note down which context their marker colour represents, e.g. green = school.
	• If possible, take photographs of the finished sheets and share them with students.
		• For example:
		  • A group has been allocated the context 'School'. They receive the piece of paper with the title "Instant messaging" and add a key at the top saying that their green pen = school.
		• On the paper, they note down:
		  • Benefits: useful in an urgent situation, for example if a teacher needs to be contacted quickly
		  • Limitations: cannot be used to send large amounts of information or documents
		• When the 5 minutes are up, they pass this sheet on to another group, and receive the next sheet titled 'Social media'.
		• See the example illustration in the Student's Book.

3	• Ask students to complete (Page 80) Task A, 'Choose and justify a communication method'. For the scenarios given, students need to choose an appropriate communication method and write down the reasons why they chose that method. • Talk through the scenarios with the class and ask students to volunteer to explain their choices.	• For each of the scenarios it is possible to justify more than one of the different communication options.

Activity	Instruction	Guidance
4	• Ask students to complete Build 5 (📖 Page 109).	• Students should not feel that they need to cover everything from the lesson in their booklets. The booklet is their story of how they would get a message to their virtual pen pal, so they may wish to focus on just one communication method of their choice.
5	• Reflection: Ask students to complete (📓 Page 81) 'Reflection'. Students reflect on their preferred method of online communication.	• Miscommunications can often happen online because text-based communication lacks other cues such as facial expressions and tone, which are present when speaking face to face. This theme is referenced in the optional homework task.
6	**Homework (Optional)** (📓 Page 81) • Online or offline? Students consider the benefits and drawbacks of online vs offline communication.	• In this situation, offline means any method of communication that does not require a device and a connection; for example, talking face to face or sending something via mail.

Build on this: Can you think of any other online communication media that were not discussed? Examples include internet chat rooms, bulletin boards and business communication methods such as Microsoft Teams or Slack. Find out the benefits and drawbacks of an additional online communication method.

Chapter 6: Lesson 5 Data transmission

Student's Book page 110			
Learning Objective/s	• Explain how an echo check works • Define compression and describe why it is needed		
New keywords	**Packet** – a chunk of data being transmitted across a network, e.g. the internet **Compression** – reducing the size of a file without affecting its contents **Echo check** – a test to see whether data has been transmitted correctly, which repeats the data back to the sender and checks for differences **Run length encoding (RLE)** – a compression method that allows files with repeating patterns to potentially be stored more efficiently **Transmission error** – a problem that occurs when data is sent across a network		
You will need	• 'Echo check messages' worksheet • Coloured pencils or pens	**Key pedagogy**	Unplugged

Activity	Instruction	Guidance
1	• Remind students that they have already learned that cables are sometimes affected by interference, which means that the data doesn't get through correctly. One way to test whether the data is correct is to use an echo check. • An echo check just means that the receiving computer repeats back the message (Page 110). • Organise students into pairs for the role-play activity. Give one student from each pair one of the message slips from the 'Echo check messages' worksheet. This student should verbally transmit the message to their partner who writes it down in (Page 82) Task A, 'What was the message?', and then the partner reads the message back to check it is correct.	• An echo check is a fundamental form of transmission check and is not the only way that computers can verify data that has been transmitted correctly (see 'Build on this' section). • Once you have finished the activity, you could ask students how easy they found the task, and whether they think this is an effective method a) for a human and b) for a computer. • Humans often use this method of verification when sharing information with each other; for example, two people performing a manual stock check in a business might have one person call out the stock levels and another repeat what was said before writing it down.
2	• Ask: 'If there was less data to transmit, would this mean fewer errors?' • Compression means to reduce the size of a file while keeping the contents the same (Page 111).	• Students may have seen vacuum packed items such as food or bedding before. You could draw an analogy with compression – it is still the same product; it has just been compressed to be smaller in size.
3	• Demonstrate run-length encoding on the board, using the example and method (Page 111). • Ask students to complete (Page 82) Task B, 'Run-length encoding'. Students use the same message from the previous exercise and write down the run-length encoded version. • Use Role play 2 (Page 112). Students transmit the same number – is this method more efficient? • Ask students to calculate how much shorter their compressed message was. • Message 6 can be compressed the most (49 down to 36 characters). Messages 2, 3 and 5 actually increase in size when they are 'compressed'! Why do students think this is?	• You could ask students to keep a record of how many mistakes were made in each version of the task, i.e. compressed and uncompressed transmission. It is important to avoid the misconception that computers make 'mistakes' in the same way that humans do. However, it might be interesting to see whether having less data to transmit did result in fewer human errors. • Students may be able to draw conclusions about the type of data that might be best compressed using run-length encoding, i.e. data that has a lot of the same letter repeating in a row. There are lots of other different compression methods.

Activity	Instruction	Guidance
4	• Ask students to complete (Page 83) Task C,' Decode the message'. Students decode the run-length encoded message into an image, by choosing a different colour for each letter in the message, and colouring in the grid provided.	• Students should use the key to make sure that they always use the same colour every time a particular letter occurs in the message. It does not matter which colours are used to complete the image. • Students should start with the top left square and work their way to the right. When they reach the edge of the square, they should move down onto the next line and start again from the left. • If a run is longer than one line (e.g. W8) then students should simply continue on with the run on the next line until the correct number of boxes have been coloured.
5	• Reflection: Ask students to complete (Page 83) 'Reflection'. Students reflect on their paired work and the most interesting part of the lesson.	
6	• Use Build 6 (Page 112). In the project book, students should add information about echo checks and compression.	• This is the last lesson where students will be adding content to their booklet. There will be opportunities during the final lesson to edit and add finishing touches to anything not yet completed. • If devices are available, students could take photographs of their pixel pictures to feature in their booklets alongside their run-length encoding.
	Homework (Optional) (Page 84) • Students colour a pixel picture and then write down the run-length encoding for their picture.	• This task gives students the opportunity to both be creative, making a picture, and consolidate their understanding of the compression technique taught in the lesson.

Build on this: Students could predict and/or check whether the image they draw from the run-length encoded version of the message is identical to the image drawn from the original copy of the message.

The image is identical because the compression method is *lossless*, i.e. none of the data is lost. Some image compression methods such as JPG are *lossy*, which means that, to compress the data, some of it is approximated. Sometimes you can get slightly fuzzy areas or marks on images that have been compressed, because fine details of the image have been lost in the compression process.

Chapter 6: Lesson 6 Finish your booklet

Student's Book page 113	
Learning Objective/s	• Showcase and explain a piece of work to peers
Key pedagogy	Showcase

Activity	Instruction	Guidance
1	• Use Build 7 (Page 113). Allow students time to finish any remaining parts of their booklet.	• Give students a time limit for when they have to finish, and regular reminders of how long they have left.
2	• Run the showcase (Page 113). Students gather in groups of approximately four. Each student should talk about one specific part of their work that they are the most proud of and explain why.	• The aim here is for students to talk about their work confidently to their peers. Choosing a part of the work they are particularly proud of will hopefully make this process easier and less daunting for students who are not used to showcasing their work in this way.
3	• Ask students to complete (Page 85) Task A, 'Feedback'. Split the groups of four in half, so that each student has a partner. Ask each pair of students to swap Workbooks and project booklets. They should peer evaluate using the assessment checklist provided. • Students then swap back and read the peer feedback they were given.	• The aim of this task is to encourage students to evaluate the work of another student. Often students are asked to evaluate their own work and they may be either over- or under-generous with their evaluation. • Encourage students to be honest but polite with their feedback. If they think the work could be improved, they should make feasible suggestions.
4	• Reflection: Ask students to complete (Page 86) 'Reflection'. Students reflect on their work during this chapter, focusing on how well they met the project brief and what they found the most useful.	• Students have been working towards project briefs in all chapters, but this chapter is particularly explicit about the objectives they should cover.

Activity	Instruction	Guidance
	Homework (Optional) (Page 87) • Students draw a picture that includes as many things as possible that they have learned during this chapter, to use as an illustration for the front page of their booklet.	• They could take a photograph or scan the page from their Workbook to add to their finished booklet.

Build on this: You could ask students to write down a piece of positive feedback and a suggestion for improvement for each person in their group regarding their verbal presentation style. For example:

Positive: "Your explanation of why you had drawn the cables as a cross-section was really clear."

Improvement: "At times during your presentation you were looking at the booklet rather than the audience, and this meant it was hard to hear what you were saying. Try to make sure that you look at the audience when you are presenting."

Chapter 7 The power of data

Project: Library database

Chapter overview

Have you ever wondered how when you watch a TV show or buy something online, you are shown accurate recommendations for other things you might want to buy or watch? This chapter will take students on a journey, from how to find books in the library to creating their own database of books and book ratings. They will use simple methods to generate a book recommendation for another member of the class and consider how recommendations are generated in the modern world using machine learning.

Chapter 7: Lesson summary		
Lesson	**Learning Objectives**	**Summary**
1	• Evaluate the suitability of a pre-existing database for a specific purpose	Students investigate a library card cataloguing system and compare it with a modern library database. They will then produce a poster evaluating the suitability of each system.
2	• Describe and use a linear search • Explain when a searching algorithm might be used	Students will learn how to perform a linear search to find an item in a list. They will trace and modify pseudocode for a linear search.
3	• Design a data collection form for a specific purpose, including data validation • Evaluate the suitability of data that has been collected for particular purposes	Students find out about the different types of form inputs, along with some basic validation techniques. They then design a data input form for adding new library books and showcase their design to a partner as a pilot.
4	• Explain how to add validation rules to a database	Students create the data capture form they have designed using software. They add data validation to the form, and then test it by capturing data about ten books.
5	• Identify the data required for specific applications	Students think about what data they would require to be able to make a book recommendation to someone who visits the library. They then create an additional ratings table for their database and populate it collaboratively within a group.

6	• Define machine learning	Students use the ratings data they collected in the previous lesson, along with some strategies for recommendations, to generate a recommendation for another member of their group. They then discuss how machine learning can provide recommendations and is used in real-world situations such as recommending TV shows to watch.

End of chapter project	**Library database**
Example ideas that learners could design:	• A database that can recommend a new book to read based on a person's previous ratings of books

Why we are learning this	
Useful skills covered in this chapter	• Research skills – how to find books in a library
Links to future careers	• Librarian – cataloguing books • Artificial intelligence – how machine learning can use huge amounts of data to provide recommendations
Links to other subjects	• Mathematics – number bases • Modern foreign languages – how characters from different languages are encoded on a computer • English – genres, authors

Chapter 7 – The power of data

Project: Library database

Name	

	Working towards (1)	**Meets expected (2)**	**Exceeding expected (3)**	**Score**
Data capture form	• Designs a data input form for recording information about library books, such as title, author, year of publication or genre	• Creates a data capture form using suitable software that can hold information about library books, which includes at least one type of data validation, for example, a type check, length check or presence check	• Adds data validation to the data capture form, and then tests it by capturing data about ten books • Uses a range of input types in their data capture form	
Identify the data required	• Considers the data needed to be able to make a book recommendation and how that data can be collected	• Creates an additional ratings table for their library database and populates it collaboratively within a group	• Understands that the BookID primary key from the books table is being used as a foreign key in the ratings table and that one book can have many ratings	
Design a database that makes a recommendation	• Uses the ratings data they collected to generate a recommendation for another member of their group	• Uses their ratings table to generate a recommendation for another member of their group and evaluates how reliable it is	• Discusses how machine learning can provide recommendations and is used in real-world situations	

Teacher feedback	
Learner response	

Chapter 7: Lesson 1 Keeping records

Student's Book page 117			
Learning Objective/s	• Evaluate the suitability of a pre-existing database for a specific purpose		
New keywords	**Database** – an organised collection of data, usually stored on a computer **Record** – a piece of information written down in a permanent form, to be used for later reference		
You will need	• A3 paper (optional)	**Key pedagogy**	Real-world concepts

Activity	Instruction	Guidance
1	• Introduce the topic by talking about how humans have kept records throughout history (Page 117). Students may be able to think of other examples. • Use Investigate 1 (Page 118). Ask students to use a search engine to find out how a library cataloguing system using paper cards works. • Ask students to complete (Page 89) Task A, 'Label the catalogue card'. Students need to label the data on a copy of the library catalogue card shown in the Student's Book.	• There is information on card cataloguing systems freely available on the web, for example: (Website C7L1R1). • Students should label the card with the different data it contains, for example author, publisher, etc. The 'call number' of a book specifies where the book can be found in the library, if the library uses the 'Dewey Decimal system'.
2	• Reflection: Ask students to complete (Page 89) 'Reflection'. Students reflect on how they feel about the two systems and think about how older people who have experienced both may feel.	• Young people have grown up with technology, so in this exercise they are asked to imagine how older people who have probably experienced both library systems may feel. You may want to ask students to share their reflections and discuss any interesting points.
3	• Use Investigate 2 (Page 118). Students then research how books are catalogued in a modern library (see guidance). • Ask students to complete (Page 90) Task B, 'Compare library systems'. Ask students to fill in their notes in the table to compare card catalogue and modern computerised library systems.	• If you are lucky enough to have a school or local public library, you could organise a visit or ask the librarian to come and talk to the class about the system they use. • If you can't visit a library, several libraries have produced materials explaining how the library works, which can be found line. Alternatively, students could research library software in a similar way to the previous task, using a search engine.

Activity	Instruction	Guidance
4	• Use Design 1 (📖 Page 118). Each student creates a poster evaluating the two systems including the benefits and drawbacks of each.	• You can either leave the choice of software students use to create the poster to them or guide them to use a particular piece of software. Alternatively, you could allow them to draw the poster on a piece of larger paper. • It is important to stress that the poster is an evaluation of the systems and not just an explanation. How suitable is each system for the purpose it was designed for? What are its benefits, and does it have any drawbacks? • Encourage students not to label the card cataloguing system as inferior simply because it has been superseded.
	Homework (Optional) (📖 Page 90) • What physical items are needed for each of the library systems?	• Answers for this task can be found in the answer document.

Build on this: Think about how people were involved in the process for each of the record keeping methods. Students could consider how technology has had an impact on the way we record and access information in general.

Chapter 7: Lesson 2 Linear search

Student's Book page 119

Learning Objective/s	• Describe and use a linear search • Explain when a searching algorithm might be used		
New keywords	**Dry run** – simulate how an algorithm will run by tracing how the value of each variable changes **Linear search** – an algorithm that checks each item in a list in turn until the search item is found, or the end of the list is reached **Searching algorithm** – a method of finding a specific item in a list **Trace table** – records how the values of variables in an algorithm change as the steps of the algorithm are followed		
You will need	• Sheets of paper or card with large numbers 1, 2, 3, 4 and two copies of 5	**Key pedagogy**	PRIMM

Activity	Instruction	Guidance
1	• Introduce the idea of a searching algorithm as being a way of finding a specific item in a list (Page 119). • Use Role play 1 (Page 119). Ask for five volunteers to role-play a linear search in a list of numbers. The five students stand in any order facing the class, holding up a card with a number on – they are the items in the list. You represent the flow of control and hold a card with a number 5. Starting with the leftmost student, work along the line asking each student in turn: 'Are you 5?' If they respond yes, the item has been found. If they respond no, move to the next student until either the item is found, or the end of the list is reached • Ask students to complete (Page 91) Task A, 'Describe a linear search in words'.	• You could relate the linear search back to finding a specific card in a given drawer, in the catalogue library system. • For the role-play activity, you could use mini whiteboards if you have them available. You can use any numbers you like on the cards but ensure that 5 is one of the numbers. The case where 5 is not in the list is considered in the 'Build on this' task. • Searching algorithms can be used on other data structures than a list, for example, a tree, linked list or a graph. Students do not need to know the definition of a data structure at this stage, so we have deliberately kept this term out of the definition to avoid confusion.
2	• Present the pseudocode for the linear search (Page 120). • Ask students to complete (Page 91) Task B, 'Answer the questions about the trace table'.	• The pseudocode uses a FOREACH loop, which has not explicitly been covered in pseudocode but should be able to be understood. If necessary, discuss line 3 with the students and mention that the end of this loop is signified on line 7 – you can tell because they are both level in terms of indentation. • Students may expect the answer to (Page 91) Task B, question 4 to be that the algorithm stops executing because the search item has been found. However, Nikesh has made a mistake in his pseudocode and the algorithm will continue to check all of the rest of the items in the list, which is a very inefficient implementation. See 'Build on this' section for information on best and worst case scenarios. • Advanced students may realise that they could swap the FOREACH loop for a WHILE loop to avoid checking every item in the list after the item is found. This is deliberately avoided in the example to keep things simple.

Activity	Instruction	Guidance
3	• Use Build 1 (Page 121). Students work in pairs to modify the pseudocode, adding a Boolean variable so that it outputs a message if the search item was not found. • Ask students to complete (Page 92) Task C, 'Record the new algorithm'.	• This activity uses the PRIMM methodology, which encourages students to modify existing code. Students at Stage 8 would not be expected to implement a linear search algorithm in code by themselves. • The key points for success in this activity are: 1. Add a Boolean variable **found = False** anywhere before line 3. This records whether the search item has been found and has a starting value of False because the search has not yet begun. 2. After the program has output "Found it" on line 5, add a line of code to set the value of the **found** variable to True. This keeps a record in the program that the search item was found. 3. After the end of the existing code, add an **IF** statement, which will check whether the **found** variable is still False and, if it is, output a message saying that the search item was not found. This statement should be at the end of the program because it should be executed only after ALL of the items have been checked.
4	• Reflection: Ask students to complete (Page 92) 'Reflection'. Students reflect on how well they could explain a linear search to someone else and think about whether they would use a linear search to find a library book.	• Linear search is a good choice in a list with very few items in it. In a very long list, it could take a long time to find something. So, finding a book in a library using a linear search would probably not be the best idea! • This may cause students to reflect that they would probably narrow down their search area first; for example, by going to the fiction section, or in a card catalogue they might go to the drawer for a particular letter. There are many other searching algorithms, some of which are extremely complicated.

Activity	Instruction	Guidance
	Homework (Optional) (Page 93) • Students are presented with a scenario relating back to the library where they have to analyse the number of comparisons and the best and worst case scenarios.	• This task introduced students into the measurement of 'number of comparisons' required in a friendly way. They may realise from question 4 that the 'worst case scenario' is always that the item they are searching for is in the list but is the last item in the list.
Build on this: To add extension to the role-play activity you could ask students to rearrange the order in which they are standing to make the best and worst case scenario for the linear search. The best case scenario is where the search item is the first item in the list, so you only need to do one comparison to find it. The worst case is when the search item is the last item in the list, or is not in the list at all, so you need to do *n* comparisons to find it, where *n* = the number items in the list.		

Chapter 7: Lesson 3 Data capture forms

Student's Book, page 112	
Learning Objective/s	• Design a data collection form for a specific purpose, including data validation • Evaluate the suitability of data that has been collected for particular purposes
New keywords	**Checkbox** – a form input that allows one or more choices to be made from a list of options **Form** – a method of collecting specific data, often to be added to a database **Length check** – checks that the length of the data typed in is within a range **Presence check** – checks whether a form field has been filled in **Radio button** – a form input that allows a single choice to be made from a list of options **Text box** – a form input that accepts string data **Type check** – checks that the data entered has the correct data type **Validation** – data typed into a form is checked to see whether it is a sensible value
Key pedagogy	Showcase

Activity	Instruction	Guidance
1	• Explain that a modern library receives new books regularly, so it needs to be able to add these books to the database. • Go over the different types of form input (Pages 122 and 123).	• Students will have seen these input types in Stage 7 when they learned to evaluate the effectiveness of data capture forms.
2	• Teach the different types of data validation (Page 124). • Ask students to complete (Page 94) Task A, 'Would it be accepted?'. Students look at the validation rules described and decide whether or not the data would be accepted. • Go over the answers to the Workbook task with the class.	• There are other more complex forms of data validation that are out of scope for this lesson; for example, checking that an email address meets the required format. • Note that each validation check performs only its stated purpose; for example, "three" is not valid data for a number type check, because although it is the name of a number, the validation is expecting either an integer or real data type. • Radio buttons have inbuilt validation because they only present the options that have been specified. It is not possible for the user to input "b" or "6" in question 2, because those options do not exist as radio buttons to select.
3	• Ask students to complete (Page 95) Task B, 'Design a data capture form'. Students design a form to collect data about new books, which must include at least one instance of all three validation rules covered: length check, presence check and type check. • Discuss with the class and decide on a subset of fields that *must* be collected on the form (this will be needed in a later lesson).	• You may wish to refer back to students' findings in Chapter 7.1, including the information they labelled on the card in Lesson 1, Task A, 'Label the catalogue. • Students could also use a book they have to hand (perhaps a school textbook or a reading book) to provide a prompt for the information they may need to store. Remember that there is other information that might also be required by the library that is not related to the content of the book; for example, the number of copies available.

Activity	Instruction	Guidance
4	• Use Showcase 1 (Page 125). Students pair up and take it in turns to role-play either a software developer presenting their form design, or a librarian giving feedback. They should offer verbal feedback on their partner's design as a 'pilot' for the form design.	• If you have a school library, you could invite the librarian to this lesson to give students feedback on their forms.
5	• Reflection: Ask students to complete (Page 96) 'Reflection'. How well did your form design pilot go? Do you need to make any changes?	• Students can note down any action points raised in their feedback during the reflection task, so that they are not forgotten before the next lesson.
	Homework (Optional) (Page 96) • Annotate the features of a data capture form.	• Students are given a data capture form and asked to label the input types and suggest validation that might be used for each input. • An answer for this task is provided in the answer document.

Build on this: There are other types of form input. A drop-down menu provides very similar functionality to radio buttons but uses up less space on the screen. Some forms use a date selection input type, which makes it easier for a user to select a year and date on a calendar type interface. Students could research and use additional input types in their form design.

Chapter 7:	Lesson 4 Data validation		
Student's Book page 126			
Learning Objective/s	• Explain how to add validation rules to a database		
New keywords	**Character** – any single letter, number, space or punctuation mark		
You will need	• Form creation software (e.g. Google Forms)	**Key pedagogy**	Explain to others

Activity	Instruction	Guidance
1	• Demonstrate how to create a form and apply validation with the form tool you are using. The example shown in the Student's Book uses Google Forms (📖 Page 126).	• Google Forms is a simple way of creating a data capture form and applying validation to the input fields. The data from the form is collated as a spreadsheet. As students only need to know about single table databases at this point in their studies, you can teach the objectives in this chapter without database software. • However, if you prefer and have the software available, students could complete this task using Microsoft Access. They would also need to create a single table database in Access to hold the data and link each form field to a database field. If you choose this method, you may need to reduce the time allocated to Build 2 (📖 Page 127) or to set (📖✏ Page 97) Task A, 'Draw and explain how you added data validation' as a homework task, as the set-up will be more involved.
2	• Use Build 2 (📖 Page 127). Students build the form they designed in the previous lesson as a digital form. As they build, they should test that the validation rules they have added work correctly; for example, if they type in some text that is too long or too short in a field that has a length check applied to it. • Ask students to complete (📖✏ Page 97) Task A, 'Draw and explain how you added data validation'. For each type of validation, students should draw a field where they used it, and explain how they added the validation, including any criteria they used.	• Students should refer back to (📖✏ Page 126) and act on any feedback they received from their partner in the previous lesson. • The Workbook task in this part of the lesson ensures that students have met the learning objective to explain how to add validation rules to a database.

Activity	Instruction	Guidance
3	• Use Build 3 (📖 Page 127). Students use the forms they have built to collect data about ten books.	• Students can find the details they need about books on a variety of searchable databases online. Some examples include: • (Website C7L4R1) • (Website C7L4R2) • Encourage students to choose books they are interested in or have read. In the next lesson they will work together with a group and amalgamate their data, so it would be good to have a variety of books to avoid repeated data. • Google Forms collects the data that students type in and adds it to a spreadsheet automatically. To view the data that students have input, from the form design page click on the 'Responses' tab, then click on 'View in sheets' and choose to create a new sheet if you are prompted to do so.
4	• Reflection: Ask students to complete (📖 Page 98) 'Reflection'. Students are asked whether their validation helped to catch any errors, and which type of validation they think is the most effective.	• Students reflect on the validation they added, and whether it helped them to catch and avoid any errors when they input the data. • It is never possible to use validation to ensure that data is correct, only that it is sensible and within specified constraints. For example, I could fill in a form and write that my name is Farah. My name is *not* Farah so this information is incorrect, but the form should accept Farah as an input as this could be someone else's name.

Activity	Instruction	Guidance
	Homework (Optional) (Page 98) • Students gain practice at writing good error messages so that someone filling in a form understands how to input the information correctly.	• It is important to provide clear error messages to explain to a user why what they typed did not meet the validation rule. This task also tests whether students fully understand and can explain what data will pass the validation checks according to the rules that have been set.

Build on this: When you create a database (as opposed to a spreadsheet) you often add validation directly to the database table itself. Here is an example of a database table called 'Books', which currently only has one field, 'copies' to track the number of copies of the book.

The copies field has a *type check* because the data type is set to integer. The phrase 'NOT NULL' (or 'NN' in the table) means that the field must not be left blank, which is a *presence check*. You can also specify additional validation checks; for example, this field only accepts values greater than zero (note that this is NOT a length check as it is not checking how many characters have been typed).

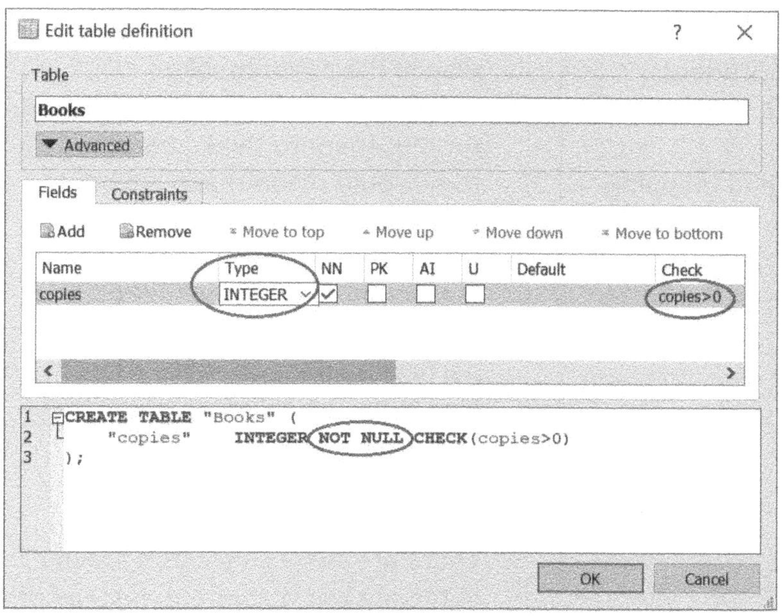

Figure 7.4.1 Example of a type check.

Chapter 7:	Lesson 5 Recommend a book		
Student's Book page 128			
Learning Objective/s	• Identify the data required for specific applications		
New keywords	**Primary key** – an assigned field in a database table that contains data that is both unique and not empty for every record **Table** – a collection of related data, stored as columns and rows		
You will need	• Spreadsheet software	**Key pedagogy**	Unplugged / Collaboration

Activity	Instruction	Guidance
1	• Use Discuss 1 (Page 128). What information would you need to collect to be able to recommend a book to someone who comes to the library? • Ask students to complete (Page 99) Task A, 'Draw a mind map'. Students draw a mind map to explore the data they might need to collect • Discuss the results of the mind mapping exercise together as a class.	• Allocate students to a group of four. They discuss how they would recommend a book and they should each note down the data they would need in the form of a mind map in their Workbooks. • If students are stuck getting started, prompt them to think about how they would recommend a book to a friend, and what information they would need to be able to do that. For example, they might know that a friend likes a particular author or a particular genre of books. They may also know which books the friend has already read, and what they thought of those books – did they like them or not? • There are some suggested answers in the answer document. Instead of discussing the results, you could draw a mind map on the board and invite students to come up and add their suggestions.

Activity	Instruction	Guidance
2	• Without looking at (Page 128) ask students to consider whether it would be a good idea to add book rating data to their existing books table, or whether this might result in some problems. • Recap the concept of a primary key, which students studied in Stage 7. • Use Build 4 (Page 129). Students are going to work in their groups to combine the data they have collected into a single table. • Use Build 5 (Page 129). Ask students to add a new field to the books table they have already created called BookID and to give each book a unique integer number. • Students should give numbers to *all* of the books that have entries in a table, across all of the tables within the group, ensuring that each book has a unique number.	• It is not a good idea to add book rating data to the books table because there could be multiple ratings per book. This would either mean 1) adding an unknown number of columns to the table to hold ratings, or 2) repeating the data for the same book multiple times, one for each rating. • Students should recall that a suitable primary key field must be unique and must not be left blank. A table can have only one primary key. • It is not a good idea to use book title for a primary key because it cannot be guaranteed to be unique – there may be more than one book with the same name. ISBN could be a suitable primary key as it is unique for each book; however, these are very long. • Students will need to share data so that it can be combined into a single file. Ask them to share a link to their data with a classmate. You can find the link by clicking 'Send' in Google Forms, and then either email the link or choose to create a short URL. • For this exercise, students will add a new integer field called BookID to their books table and give each book a consecutive number starting from 1. If there are any duplicates where students have collected data on the same book, they should only include one of the instances.

Activity	Instruction	Guidance
3	• In Build 6, each group should create a ratings table, with the format specified (📖 Page 129). • Students fill in the ratings from each person in the group in the group's single ratings table.	• Each group should only create one ratings table and use it to fill in their ratings collectively. If you are using software where it is possible to collaborate on the same document (for example, Google Sheets), group members could add their ratings simultaneously. If not, one person could be the designated typist and one person could oversee looking up the ID number of a book and asking who has read it to offer a rating. • If students have not read enough books to fill in ratings, they will still need to add ratings to be able to complete the task. They could make up some ratings, although this will obviously influence the usefulness of the finished recommendation system. Alternatively, if you find that many students are in this position you could crowd-source ratings for a few books from the whole class and write the data on the board.
4	• Reflection: Ask students to complete (📓 Page 100) 'Reflection'. Students reflect on the group work tasks in this lesson and whether they could improve the way in which they work as a group for when they are asked to do this again.	• You could take in students' Workbooks and read their reflections from this task, and then offer an anonymised summary to the class to help students understand what did and didn't work well.
	Homework (Optional) (📓 Page 100) • Students look at a recommendation system they have encountered and consider how it works.	• Students may not be able to fully articulate how the recommendation system they are describing works, because it is highly likely to use complex machine learning. However, the purpose of this task is to trigger students to think about systems they have seen in everyday life, in preparation for the last lesson in this chapter where machine learning is explicitly considered. • If you think students may need some pointers, you could initiate a class discussion to give them some ideas prior to setting the homework task.

Build on this: The BookID primary key from the books table is being used as a *foreign key* in the ratings table. If students were building a relational database using database software, as they may do in future stages, this key would be used to build a relationship between the two tables. This type of relationship is called a 'one to many' relationship, because *one* book can have *many* ratings. Relationships between tables are an important part of database design at more advanced stages of study in Computing.

Chapter 7:	Lesson 6 Machine learning		
Student's Book page 130			
Learning Objective/s	• Define machine learning		
New keywords	**Big Data** – an extremely large amount of data, which might be used to train a machine learning model **Machine learning** – an area of computer science that uses algorithms and data to create software that can improve its own performance **Training** – providing data to a machine learning algorithm so that it can recognise patterns in the data		
You will need	• Spreadsheet software	**Key pedagogy**	Collaboration

Activity	Instruction	Guidance
1	• Use Discuss 2 (Page 130). Students re-join their groups from the previous lesson and discuss how they could use their ratings data to recommend a book to Malik.	• Set the scene for this task – Malik walks into the library and tells the librarian that he has enjoyed a particular book. If the library database contains his ratings for the books he has previously read, how could you use that data to recommend a new book? • Ask students to close their Student's Books. Students then discuss with their group how they could use the ratings data they collected in the previous lesson to make a good recommendation. • If you set the homework task from the previous lesson, prompt students to consider what they wrote about how real-life examples of recommendation systems work.

Activity	Instruction	Guidance
2	• Allocate each student another member of the group to make a book recommendation for and explain the Investigate task. Students then use the two strategies specified in Investigate 3 (Page 130) to recommend a book to that person. • Ask students to complete (Page 101) Task A, 'Make a recommendation'. Students are asked to write down the process they followed and the recommendation that was generated. • If there is time, students could come up with an alternative recommendation strategy and test that.	• It may help to recap how to sort and find data within a spreadsheet, as this will be useful when collating the recommendations. • Students are simulating queries on a database during this activity, without actually using database software. Looking up the genre of the book via the BookID from the ratings table is how data would be linked in a relational database, where BookID is the primary key in the books table and a foreign key in the ratings table.
3	• Talk through the material about machine learning (Page 131). • Ask students to complete (Page 101) Task B, 'Write a definition of machine learning'. Students are asked to use what they have learned this lesson to write a definition of machine learning.	• This is only a brief activity because the area of machine learning is extremely complicated, and students are only expected to be able to define what it is at this stage. However, you could extend this task with the suggestion from the 'Build on this' section.
4	• Students now Showcase their work (Page 131): Showcase 2. Ask students to share with the class the book recommendations their strategy came up with.	• For example, a student may say: "Anika, I recommend you read <book>" and then Anika could respond with whether she thinks this is a good recommendation.
5	• Reflection: Ask students to complete (Page 102) 'Reflection'. How do they feel about machine learning?	• This reflection task invites students to think about how they feel about machine learning and its use in everyday situations. How do they feel about their data being used to train machines?
	Homework (Optional) (Page 102) • This asks students to write a book recommendation to display in the library.	• If you have visited your school library or had a visit from a librarian in your lessons, it might be nice to round off this chapter by asking students to write up some of the book recommendations generated by their investigations. These could then be photocopied and provided to the library to display.

Build on this: Demonstrate an application of machine learning using a real large language model (LLM); for example, ChatGPT and Google Gemini are both free to use with an account. Ask students to tell you the title of a book they have enjoyed and ask the LLM to use that information to give a book recommendation. The quality of the recommendations is usually very high.

However, note that the information provided by an LLM is not always true or accurate. For example, ChatGPT was asked to recommend books similar to a particular book, but it recommended the same book and also incorrectly stated that it was written by different authors!

Chapter 8 Create with code 3

Project: Technology at work

Chapter overview

Students will start by focusing on how technology is used in a variety of workplaces, both human controlled and autonomous. They will learn how to use the radio communication facilities of the BBC micro:bit, and then design a program that uses two communicating micro:bits to solve a problem in the workplace. They will dry run the algorithm, devise a test plan and test their program with test data.

Chapter 8: Lesson summary		
Lesson	**Learning Objectives**	**Summary**
1	• Explain the evolving role of digital technology in the workplace	Students will investigate the use of digital technology in the workplace, working in a group to create a presentation focusing on a specific workplace.
2	• Identify uses for autonomous programming and AI in robotics	This lesson focuses on autonomous programming and the workplace situations in which autonomous robots are used.
3	• Develop a program that uses two or more interacting physical devices	Students learn about radio communication and how it can be used to communicate between two micro:bit devices. They write a simple program, and then design a device using radio communication that can be used in a workplace.
4	• Dry run an algorithm using suitable test data	Students will design an algorithm using pseudocode to make their devices work, and then learn how to dry run an algorithm.
5	• Develop and use test plans to test a program	In pairs, students will build the device they have designed. Once it is built, they will create a test plan with sample test data.
6	• Explain why it is sensible to use a range of test data	To complete the topic, students will use the test plan they devised to test the device. They will then demonstrate what they have created to another group.

End of chapter project	Technology at work
Example ideas that learners could design:	• A detector to scan items leaving a shop • A checkpoint scanner to track delivery vehicles arriving • A device to locate an item in a warehouse

Why we are learning this	
Useful skills covered in this chapter	• How radio communication works • Two different methods for testing a program – dry run and test plan
Links to future careers	• Consideration of the technology required for a variety of careers • Software engineering – how to test a system
Links to other subjects	• Physics – radio waves

Chapter 8 – Create with code 3

Project: Technology at work

Name	

	Working towards (1)	Meets expected (2)	Exceeding expected (3)	Score
Design a device	• Understands the use of digital technology in the workplace • Selects a suitable scenario for their device and describes how it works	• Writes a simple program that involves interaction between two micro:bits that can be used in a specific workplace	• Is creative in their design and uses a range of functions on the micro:bit	
Build the device	• Designs an algorithm using pseudocode to make their device work	• Dry runs the algorithm and makes amends to the code • Builds a device that allows two micro:bits to communicate with each other via a radio signal	• Uses a range of features of the micro:bit; for example, the buttons or the LED matrix	
Test the device	• Creates a test plan with sample test data	• Uses a test plan with a range of data to test their device and then make improvements as a result • Demonstrates what they have created to another group and shows that it works as planned	• Uses extreme data in their test plan to check that their device works correctly	

Teacher feedback	
Learner response	

Chapter 8: Lesson 1 Digital technology in the workplace

Student's Book page 134			
Learning Objective/s	• Explain the evolving role of digital technology in the workplace		
New keywords	**Barcode** – a series of stripes that contain data and can be scanned by a barcode reader device **Digital technology** – any device or system that creates, stores or uses data **Radio Frequency Identification (RFID) tag** – contains data that can be read via a radio signal when the tag is close to a reader		
You will need	• Presentation software	**Key pedagogy**	Collaboration

Activity	Instruction	Guidance
1	• Use the Student's Book to define the term digital technology (Page 134). • Use Discuss 1 (Page 134). • Talk through the examples of how digital technology is used in different workplaces (Pages 134 and 135).	• Each workplace only has a single example of the use of technology given, but there are many others for the workplaces listed.
2	• Choose 5–6 different workplaces and allocate each student a workplace to focus on. Then ask students to complete (Page 104) Task A, 'Mind map'. Each student creates a mind map of ways in which technology is used in their allocated workplace.	• You could use the five workplaces discussed in the Student's Book but if your class discussion yielded some other interesting workplaces or there is a particular type of workplace that you know students will be familiar with, you can substitute or add those too. • The students focusing on each workplace will be brought together to form a group later, so make sure you have a sufficient variety, otherwise the groups will be too large.
3	• Talk through the material on hardware devices used in the workplace (Pages 136 and 137). • Ask students to complete (Page 105) Task B, 'Hardware devices in the workplace'. Students choose and describe two hardware devices and explain what each would be used for in their allocated workplace.	• It is not possible to provide an exhaustive list as the hardware required might be quite specialised to a particular industry. • Remind students that the focus of the lesson is *digital technology* in the workplace. This does not include machines that do not use data, e.g. a bread slicer or a vacuum cleaner.

Activity	Instruction	Guidance
4	• Use Build 1 (📖 Page 137). Bring together all the students who focused on the same workplace to form a group. There will a group for each workplace. • Then ask each group to create a presentation about digital technology in the workplace they have been studying. They should use a suitable form of presentation software.	• Ensure that the groups are small enough so that each person can contribute. Students could discuss their findings and then each work on a particular use of technology in the given workplace, before bringing together their slides. This approach avoids the situation where some students contribute more to the task than others. • Note that students do not need to present their work at this point – more will be added to the presentations during the following lesson.
5	• Reflection: Ask students to complete (📓 Page 105) 'Reflection'. Students reflect on what they have learned during the lesson and think about the technology that might be present in the workplace when they are old enough to work.	• This activity is more of a forward reflection – they are using what they have learned in the lesson, but then reflecting on what this might mean for them in the future.
	Homework (Optional) (📓 Page 106) • Students ask an adult about the technology that is used in their workplace.	• Students could ask their parents or, older siblings, friends, activity instructors, etc.

Build on this: Students could think about the advantages and disadvantages of each of the technologies. For example, self-scanners in shops mean that you can enter a shop and purchase what you want without ever having to interact with a human. Is this an advantage or a disadvantage, or both? It means that you can complete your shopping trip quickly and efficiently and there is less waiting, but it also means that fewer humans need to be employed and some people may miss the human contact.

Chapter 8:	Lesson 2 Autonomous programming and AI
Student's Book page 138	
Learning Objective/s	• Identify uses for autonomous programming and AI in robotics

New keywords	**Artificial intelligence (AI)** – the simulation of human intelligence by a machine **Automatic** – the ability of a machine to perform a task without input from a human **Autonomous** – the ability of a computer to perform a task without input from a human, but including the ability to react to circumstances as it finds them		
You will need	• Presentation software	**Key pedagogy**	Showcase

Activity	Instruction	Guidance
1	• Talk through the material about autonomous programming (Page 138). • Compare this with the material about automatic vs autonomous machines (Page 139).	• It is a common misconception that all machines used in production are autonomous. True autonomous robots involve some element of being able to make decisions based on inputs from their surroundings.
2	• Ask students to complete (Page 139) Task A, 'Is it autonomous?'. Students fill in the table by ticking the boxes they think apply, and use the final column to justify their answer. • Talk through the answers as a class.	• The answers are provided. • You may wish to talk through what each of the robotic systems described does before students are asked to complete the task.
3	• Reflection: Ask students to complete (Page 107) 'Reflection'. Did the students have any misconceptions about automation?	• At this point, it might be good to allow students to reflect on their own potential misconceptions about automation, to avoid carrying these forwards into the following task. You could ask willing members of the class to share their thoughts.
4	• Link back to previous lesson. Ask students: 'Are any of the uses of technology you covered in your presentation autonomous?' • Ask students to rejoin their group from the previous lesson for Build 2 (Page 139). For each use of technology, add one additional slide describing whether or not the technology uses automation.	• This task should result in each student adding one additional slide. Students could each take the same slide they created in the previous lesson, or you could ask them to each take a different student's slide to provide extra challenge.

Activity	Instruction	Guidance
5	• Run Showcase 1 (📖 Page 139). Each group showcases their presentation to the class. Students who are watching the presentations should be asked to complete (📓 Page 108) Task B, 'Automated or not?' while they watch.	• This exercise ensures that students are focused on what their classmates are saying during the showcase, and it also ensures that they all have some notes on the workplaces they did not focus on for their own presentation. • You could ask for a quick show of hands after each slide of the presentation to assess whether students think that it involves automation.
	Homework (Optional) (📓 Page 109) • Students are asked to research a use of AI in the workplace.	• Students could take this lesson further by investigating how AI is used in the workplace. They will have heard of some uses of AI from previous chapters, but not necessarily have connected these with how they might be used in a work context.

Build on this: Students could find out about AI and machine learning and its impact on the workplace. For example, machine learning is now used to help people to write all sorts of work-related documents, such as reports, and even to write program code. Large Language Models (LLM) such as ChatGPT can generate business ideas and mock up data. What ethical issues might there be with using a LLM in this way?

Chapter 8: Lesson 3 – Radio signals

Student's Book page 140

Learning Objective/s	• Develop a program that uses two or more interacting physical devices		
New keywords	**Radio signal** – a signal within a specific frequency band, used to transmit data		
You will need	• One micro:bit device per student. If these are unavailable, the lesson can still be completed using the online micro:bit emulator (Website C8L3R1). • Two micro:bit devices, pre-loaded with the demo program.	**Key pedagogy**	Get creative

Activity	Instruction	Guidance
1	• Introduce the micro:bit and explain how it works (📖 Page 140). • Open up the MakeCode editor and highlight the three key areas: the micro:bit simulator, the available blocks and the area where you write your code.	• Students may have used a micro:bit before if they have followed this course since Stage 3. • If you have a real micro:bit to hand, you may want to pass it around the class for students to handle and look at while you are explaining. • The MakeCode editor for the micro:bit can be found at (Website C8L3R2)
2	• Talk through the material on radio signals (📖 Page 140). • Demonstrate the sample program for radio communication between two micro:bits. When you press a button on one micro:bit, it displays the letter of the button you pressed on the LED matrix of the other device. Create the project in makecode.org and use it to demonstrate how the program works.	• Students do not need to know specific details about radio signals other than that they can be used to communicate data between two different devices – in this case, the devices are micro:bits but other devices could include a two way radio or a wireless microphone. • The demo program can be found in the Answers document. You should pre-create this program in the micro:bit MakeCode editor and load a copy of the program onto two separate devices.
3	• Divide students into pairs for Build 3 (📖 Pages 141 and 142). • Each pair should write a program to send a message from one micro:bit to the other when a button is pressed. The instructions for this task are given (📖 Pages 141 and 142).	• A micro:bit uses 1 byte (8 bits) to store the radio frequency group. There are 256 ways to organise the 8 bits so starting from 00000000 to 11111111 gives the values 0 to 255. This will ensure that each program does not interact with signals sent by other students in the class.

4	• Talk through the project brief and then ask students to use Discuss 2 (📖 Page 143). With their partner, students discuss how they could use the radio signal capabilities of the micro:bit to create a system to be used in the workplace.
	• Ask students to complete (📓 Page 110) Task A, 'Note down your ideas'.
		• Students can use other features of the micro:bit in addition to the radio signal functionality. You can let students explore the available blocks in the MakeCode editor while they think about their plan. Depending on which version of the micro:bit you have, not all features will be available, so check this beforehand. Some suitable features to recommend using include:
		  • LED matrix
		  • Input buttons
		  • Movement sensor (the micro:bit can respond to being shaken, tilted or turned
		  • Music – if you are using a micro:bit v1 you will need to connect the micro:bit to a speaker or headphones to hear it. See (Website C8L3R3) for more details.
		• Allow students to think of their own ideas in pairs and circulate among the class listening to their thoughts. You can prompt students who may be stuck by asking questions such as:
		  • Can you remember any situations where RFID tags might be used? (e.g. to mark valuable goods, to scan library books)
		  • Could you use one of these ideas to design a new kind of system for a different workplace?
		• It is possible for students to use 'in range' and 'out of range' as a rough measure of proximity. The approximate maximum range for the signal is 70 metres, but you can use the "radio set transmit power" block in Radio > more... to change the strength of the signal transmitted.
		• Some students may want to be more precise than this. See the 'Build on this' section for more details about using the signal strength to determine how close together two micro:bits are.

Activity	Instruction	Guidance
5	• Ask students to complete (Page 110) Task B, 'Design a system for use in the workplace'. Ask each pair of students to design a system for use in the workplace, and fill in the details of the system they have designed in their Workbooks.	• The instructions students provide during this task do not need to be comprehensive at this stage, as this is just a plan. They will be planning their algorithm in more detail in pseudocode during the following lesson.
6	• Reflection: Ask students to complete (Page 111) 'Reflection'. Students are asked to reflect on working with a partner.	• Students think about how they can work well as a team on this project.
	Homework (Optional) (Page 111) • How are radio signals used in different situations?	• This task encourages students to consider how the RFID technology they are emulating using a micro:bit works in a real-life scenario. This may help if they are struggling to formulate a concrete idea for their project.

Build on this: Students could investigate the signal strength block and use it to determine how close together two micro:bit devices are. There is a project on the micro:bit website (Website C8L3R4) that describes the steps and gives example code.

Chapter 8: Lesson 4 Design a workplace system

Student's Book page 144

Learning Objective/s	• Dry run an algorithm using suitable test data		
New keywords	**Event-driven programming** – a style of programming where all code is written to respond to events, for example when a button is pressed **Trigger** – cause to happen		
You will need	• One micro:bit per student • Access to MakeCode editor (Website C8L4R1)	**Key pedagogy**	Get creative

Activity	Instruction	Guidance
1	• Remind students about the steps they wrote in Lesson 3, (Page 110) Task B 'Design a system for use in the workplace'. • Ask students to complete (Page 144) Task A, 'Divide up the tasks'. Ask each pair of students to divide up the steps into tasks that need to be accomplished by each micro:bit. • An example is given that uses the same scenario of a restaurant ordering system as in Lesson 3 (Page 144).	• Both micro:bits may need to be loaded with the same code – this is fine. • Use this stage to check the scope of what students are planning – if there are lots of tasks in the box, they may not be able to accomplish everything in the time allowed.
2	• Ask students to complete (Page 112) Task B, 'Write the pseudocode'. Once they have successfully divided the tasks, each student writes pseudocode for the tasks their micro:bit needs to achieve. • There is some exemplar pseudocode continuing the previous example (Page 145).	• Each student should take charge of writing the code for one of the micro:bits. It may be the case that both micro:bits can be loaded with the same code, in which case the students could either each come up with an algorithm in pseudocode, or they could collaborate, but both note it down in their Workbooks.
3	• Remind students about how to dry run an algorithm (see (Page 145) for an exemplar). • Ask students to complete (Page 113) Task C, 'Dry run your pseudocode'. Ask students to draw a trace table and test one part of their pseudocode with a sample (valid) input.	• Students have dry run some pseudocode before in Chapter 7.2. However, it is worth a recap during this lesson as this time parts of the code will only execute in response to a particular trigger, such as pressing a button, or a radio message being received. The program will run the initial setup (in the exemplar this is lines 1–3) and then will wait for an event such as a button press to occur. In the trace table shown in the Student's Book, the blank line is where the input occurs – A is pressed. This triggers the block of code for 'WHEN button A IS PRESSED' beginning from line 5. • They do not need to dry run every part of the planned algorithm or use more than one test input. Further tests will be completed during the following lessons.

4	• Reflection: Ask students to complete (113) 'Reflection'. Students reflect on whether the dry run raised any errors, and how useful they found this process.	• If any errors are raised during the dry run process, students should note these down and alter their plan accordingly.
5	• During Build 4 students can start to build the program they have planned, using the MakeCode editor (Page 145).	• Direct students to begin their code, using the pseudocode plan they have created. There will probably not be much time left during this lesson, but more time is allocated next in the lesson.
	Homework (Optional) (Page 114) • Dry run the algorithm provided and fill in the table with the output.	• Students practise dry running some pre-written code relating to the chef/customer scenario running through this chapter.

Build on this: During this lesson, students may appreciate that code for the micro:bit runs in a different way to the Python code they wrote during previous chapters. Event-driven programming means that blocks of code execute when certain events/triggers occur; for example, a button being pressed. Contrast this to Python code, which runs line by line, unless some code causes a modification to the flow of control. You could ask them whether they think Scratch is event-driven too and justify their answer. It is, and there are lots of comparisons that could be made, for example "when flag clicked" in Scratch is similar to "When program starts" in MakeCode.

Chapter 8: Lesson 5 Test plans and test data

Student's Book page 146

Learning Objective/s	• Develop and use test plans to test a program		
New keywords	**Valid data** – data that should be accepted by a program **Invalid data** – data that should *not* be accepted by a program		
You will need	• One micro:bit per student	**Key pedagogy**	Collaboration

Activity	Instruction	Guidance
1	• Highlight the two different types of test data: valid and invalid (📖 Page 146). • Use Discuss 3 (📖 Page 146).	• Valid data is also sometimes called normal data. Invalid data is also sometimes called erroneous data. • An answer to the discussion task is provided. • You could link back to Chapter 7.3 where students looked at data capture forms. This is not a form, but data is still entered into the program via the inputs available, for example pressing a button.
2	• During Build 5, students continue building the program with their partner using MakeCode (📖 Page 146).	• Students may need to test their programs as they go along in an ad-hoc way. This is fine, and do not discourage them from testing as they go. The test plan is a more formal way of confirming the functionality of the system once it has been built.
3	• Once they have finished developing their programs, students need to draw up a test plan to test them. Students' test plans should include at least one test that uses invalid data. Ask them to complete (📓 Page 115) Task A, 'Create a test plan'.	• Students created a test plan to test a program in Stage 7. • Emphasise to students the importance of being specific with their language when writing a test plan. For example, in the output column: ✗ It works ✓ A smiley face is displayed on the LED matrix
4	• Reflection: Ask students to complete (📓 Page 116) 'Reflection'. reflect on the two types of testing they have experienced in this chapter – dry runs and test plans.	• Real software is tested in several different ways. Dry running an algorithm is a good way to test the logic of a piece of pseudocode. Some systems where safety is critical will have their program code tested extensively line by line because errors could cause damage or loss of life. Other systems may only be tested via a test plan, often with multiple iterations of testing as bugs are found and resolved.

Activity	Instruction	Guidance
	Homework (Optional) (📖✏ Page 117) • Students look at the example test plan and rate each test.	• This task gives students the opportunity to evaluate an example test plan and suggest improvements. The task assesses whether they have understood what is meant by being specific with their use of language when describing tests.

Build on this: There is another type of data that should be tested, called extreme data. This is data that is only just either acceptable or unacceptable, to test that the program's conditions have been written correctly. In the order number example where the restaurant only allows nine simultaneous orders, valid examples of extreme data would be 1 and 9 because they are the smallest and largest acceptable numbers, and invalid examples of extreme data would be 0 and 10 because they fall just outside the range that should be accepted.

Chapter 8: Lesson 6 Test your program

Student's Book page 148

Learning Objective/s	• Explain why it is sensible to use a range of test data		
New keywords	**Pass** – the expected outcome of the test happens when the test is performed **Fail** – the expected outcome of the test does not happen		
You will need	• One micro:bit per student	**Key pedagogy**	Showcase

Activity	Instruction	Guidance
1	• Introduce the Investigate task and direct students to use the test plan they drew up in the previous lesson to test their program and record the results (📖✏ Page 115).	• This lesson starts with a task because there is a lot to fit in. You may want to do a brief introduction to the lesson and remind students about the test plan they drew up. • If students have been testing so far mostly using the MakeCode emulator, it would be a good idea for them to load their programs onto the real micro:bits for testing at the start of this lesson.

Activity	Instruction	Guidance
2	• Reflection: Ask students to complete (Page 118) 'Reflection'. Students reflect on whether there were any problems with their test plan.	• Depending on how long ago the previous lesson was, students may have forgotten what their intentions were with the test plan, and some of their instructions may not have been sufficiently clear. This is a good opportunity for them to reflect on their test plan and how they could improve their skills if asked to write one again.
3	• Explain that during Build 6 students will act on the results of their testing (Page 148). Give each pair some time to make changes to their program following the testing, before they showcase the program to another pair.	• For some of the pairs, all of their tests may have passed so they may feel that there is nothing to do at this point. Try to visit each pair and, if necessary, see whether you can suggest an improvement to their system. You could also use this time to assess how well the students have met the project brief and how innovative they have been.
4	• Use Showcase 2 (Page 149). Ask each pair to join up with another pair. Each pair should demonstrate their program and explain its function in the workplace. • Ask students to complete (Page 118) Task A, 'Watch the demonstration'. The students who are watching the demonstration should fill in the details of the other group's system in their Workbooks. • The students viewing the presentation should offer verbal feedback.	• Encourage students to ask questions of the other group if necessary. The prompts in (Page 118) Task A should make sure that the presenting group explains their device clearly enough so that the watching group can fill in the answers.

Activity	Instruction	Guidance
	Homework (Optional) (Page 119) • Ask students to research an example of when a real-life system went wrong.	• You can either allow students to find their own example or provide them with some systems to find out more about. Some examples of high-profile or particularly serious software failures include: • Ariane 5 rocket test flight • Therac-25 radiotherapy machine* • London ambulance despatch system* • NASA Mars climate orbiter *These software failures caused injury and loss of life, which may upset some students.

Build on this: Once groups have finished presenting to each other, you could ask each pair to look at the test plan that the other group wrote. They should write an additional test for the other group's program. This task tests whether the viewers understood how the showcased system should work.

Social media post

shoeshoee

07,627 views

shoeshoee

Biodegradable shoes made of 97% plant products.

Awards

Imagine you are a judge awarding prizes to other students in your class for their work. Who would you award the following prizes to?

	**The "I felt like I was actually there" award** Awarded to: ________________ Reason:
	**Best use of multimedia** Awarded to: ________________ Reason:
	**Best overall tour** Awarded to: ________________ Reason:
	**Student's own choice of award:** ________________________ Awarded to: ________________ Reason:

Quiz feedback

Who made the quiz?

Two things I thought were good about the quiz:

★

★

The quiz would be even better if:

•

Who made the quiz?

Two things I thought were good about the quiz:

★

★

The quiz would be even better if:

•

Who made the quiz?

Two things I thought were good about the quiz:

★

★

The quiz would be even better if:

•

ASCII values for common characters

Character	Denary	Binary	Character	Denary	Binary
Space	32		U	85	
.	46		V	86	
0	48		W	87	
1	49		X	88	
2	25		Y	89	
3	51		Z	90	
4	52		a	97	
5	53		b	98	
6	54		c	99	
7	55		d	100	
8	56		e	101	
9	57		f	102	
A	65		g	103	
B	66		h	104	
C	67		i	105	
D	68		j	106	
E	69		k	107	
F	70		l	108	
G	71		m	109	
H	72		n	110	
I	73		o	111	
J	74		p	112	
K	75		q	113	
L	76		r	114	
M	77		s	115	
N	78		t	116	
O	79		u	117	
P	80		v	118	
Q	81		w	119	
R	82		x	120	
S	83		y	121	
T	84		z	122	

Unicode

You may have noticed that there is a major flaw in ASCII. It was invented in America, which means that it can only be used for languages that use the English alphabet. What about other languages, which use a different alphabet and different characters?

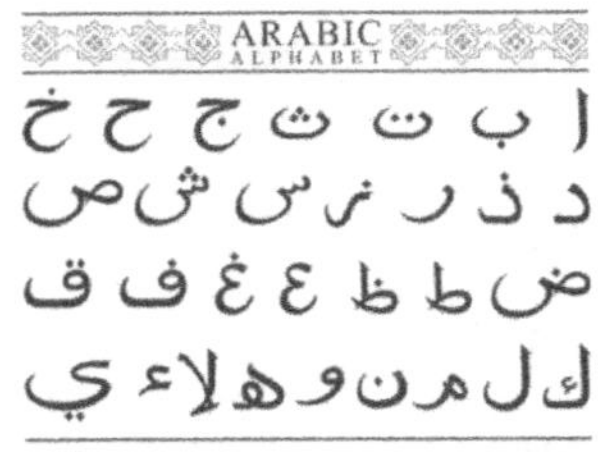

Figure 6.3.1 Arabic alphabet.

To solve this issue, a different format called Unicode was created, which has superseded ASCII, although the Unicode values of characters from the Latin alphabet are the same. Unicode uses between 1 and 4 bytes to represent each character. As of Unicode 15.1 there are 149,813 characters, including hundreds of emojis, but since there are four bytes (64 bits) available there is room for potentially 4,294,967,296 different characters to be represented!

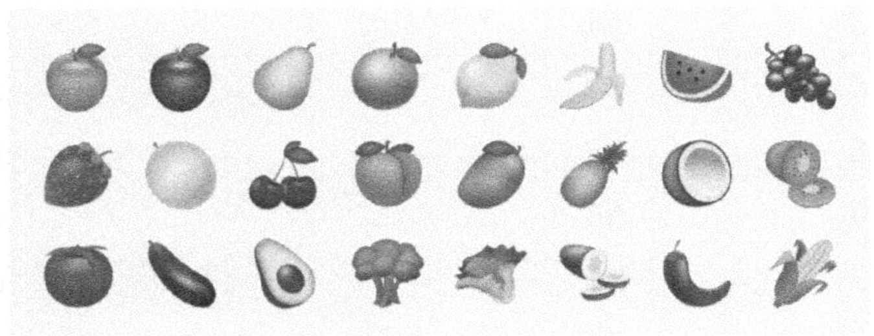

Figure 6.3.2 Emojis.

Choose a character that is not included in ASCII, for example a character in another language, or an emoji. Find out the value of that character in Unicode.

Troubleshooting tool test scenarios

Scenario 1

Your laptop is running very slowly and you are not sure why. Programs such as your internet browser or word processor open up OK, and you do have anti-malware software installed. However, when you try to connect to the Wi-Fi and browse websites on the internet, the content takes a very long time to load.

Scenario 2

You have a lot of files on your computer, and they are all organised neatly into folders so that you can find what you need easily. However, your computer keeps saying that there is no more space available. You can't delete any files because you need them all.

Scenario 3

You work in an office and you have been asked to edit a document produced by another department. You have looked for the file in the location provided to you by that department and you can open it but it won't let you type anything. Your boss keeps asking you to get the work done as soon as possible, as colleagues are complaining.

Scenario 4

You are trying to make a leaflet in a desktop publishing program for an upcoming business fair. You are using the internet to find good images to use. Your colleagues have sent you the text to go in the leaflet as a word-processed document, and you are currently on a video call with your boss so that she can give you instructions on what to do. Unfortunately, the computer is running so slowly that you can't get anything done.

Scenario 5

You work in an office on a desktop computer. When you left the office last night everything was working OK, but this morning you can't seem to access anything on the internet. You have looked at all of the cables in the back of the computer but you don't really understand what they do. Your computer is otherwise working fine.

Firewall rules

Any traffic not specifically mentioned in the table is considered to be **blocked** in this example.

Direction	To	From	Port	Action
In	Trusted network	*	22	Block and send message "Connection rejected"
In	Trusted network	115.24.3.31	*	Block
In	Trusted network	192.168.3.76	80, 443	Allow
In	Trusted network	64.63.2.21	300	Allow
Out	*	Trusted network	80	Block
Out	115.24.3.31	Trusted network	443	Block
Out	64.63.2.21	Trusted network	300	Allow
Out	192.168.3.76	Trusted network	443	Allow

Note: * is a 'wildcard' which means "anything"

Echo check messages

Print out this sheet and cut it up into individual slips for students to use during the echo check exercise. There are six different possible messages to transmit. Ignore any line breaks in the message and continue as if the whole message were part of a single line.

Message 1 WBBWBBWBRRBRRBRRRRRRRRRRRRRRBRRRR RBWBRRRBWWWBRBWW	**Message 2** WWGWWGWWGBGGBGWWGGGGWWGGRRG WGGGGGGWGGGGGGWGWGWWGW
Message 3 WBWWWBWBYBWBYBWBWWGBWWGWGGW WWGWGWWWWWGGWWWWWWGWWW	**Message 4** WYYYYWWYBYYBWYYYYYYWYRYYRYWYYRRYY WWYYYYWWWWWWWWWW
Message 5 WWWWRRWWWWWRWWWWWBWRBWWWBWR BWWWBLLBWWWBLLBWWWBBBBWW	**Message 6** WWWPWWWWWWPPRWWWPRPPPWWWOOO WWWWOOOWWWWOOOWWWWWOWWW

Message 1 WBBWBBWBRRBRRBRRRRRRRRRRRRRRBRRRR RBWBRRRBWWWBRBWW	**Message 2** WWGWWGWWGBGGBGWWGGGGWWGGRRG WGGGGGGWGGGGGGWGWGWWGW
Message 3 WBWWWBWBYBWBYBWBWWGBWWGWGGW WWGWGWWWWWGGWWWWWWGWWW	**Message 4** WYYYYWWYBYYBWYYYYYYWYRYYRYWYYRRYY WWYYYYWWWWWWWWWW
Message 5 WWWWRRWWWWWRWWWWWBWRBWWWBWR BWWWBLLBWWWBLLBWWWBBBBWW	**Message 6** WWWPWWWWWWPPRWWWPRPPPWWWOOO WWWWOOOWWWWOOOWWWWWOWWW

Chapter 1 – Our digital world

Chapter 1.1 Evaluating online information

Task A. Evaluate the URL

URL	Comments
Website C1L1A1	The website communicates securely The page is likely to be some kind of wiki It is a non-profit organisation
Website C1L1A2	The website communicates securely It is likely to be owned by a company
Website C1L1A3	The website does not communicate securely The page might be for a bicycle shop The shop is likely to be in France (Hard) The site might be for tourists who speak English, because the domain name is in English

Task B. Mind map

1.

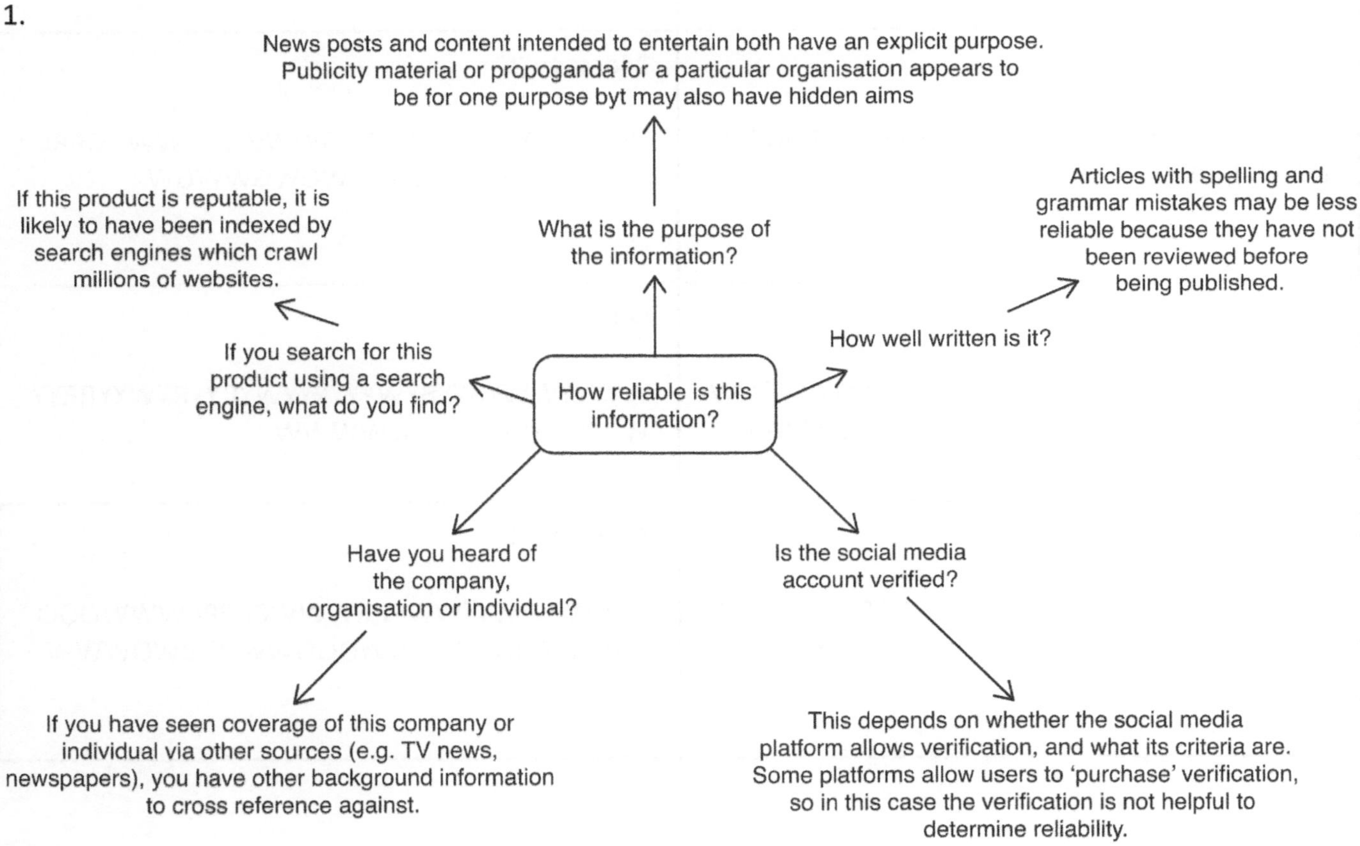

2. Answers could include:

- **Is the social media account linked to from the website of the company/individual?** This will allow you to determine whether this account is the one controlled by that company/person.
- **How many followers does the account have?** This is not an indicator of whether the content is true or false, but having very few followers may be an indicator of a copycat account.
- **When was the account created?** Often, scammers will create a fake account claiming to be a legitimate company and offering a 'too good to be true' giveaway. If the account was created very recently this is a likely indicator that it is not genuine.

Chapter 1.2 Usernames and metadata

Task A. Why choose an anonymous username?

- **Identity theft** – someone may impersonate you offline if they have your personal data, for example opening a bank account
- **Password security** – revealing personal information such as your date of birth in your username could be used to answer security questions to access your account
- **e-Safety** – if you reveal your location, someone you have met online could use it to find out where you live
  - **Reputation** – if your username includes your real name, any comments you make online are cross-linked to your real-life identity forever. This may have an impact on you in the future, for example, if a potential employer looks up your digital footprint.

Task B. What metadata is recorded?

Action	Metadata
Visit an online shopping website	• Time you visited the website • The device you were using • Items you looked at • Items you added to your cart • What country you live in (from your IP address)
Search for a route between your house and a relative's house using an online map	
Look at a website about a holiday destination	
Look at your favourite celebrity's social media page	

- They pay in US Dollars (USD)
- They visited the site on 27 October 2023 at 13:52
- They are in the United States (US)
- They speak US English
- Their username is brooke12
- They have been to this website before
- They live somewhere in Long Island, New York, USA
- Their name might be Brooke
- They might be 12 years old

- They might have been born in 2012
- They might live at number 12

Task C. Data from cookies

Note that whilst there is a password specified, this is likely to be an encrypted version of their password rather than the actual password. Passwords should not be stored in cookies as plaintext for security reasons.

Chapter 1.3 Data mining

Task A. Why is metadata useful?

Metadata	Why is it useful to a company?
Device used to visit the page	• Reveals information about the customer's income e.g. is it a top of the range device or a cheaper one? • Gives information about how they browse, e.g. may use a laptop at work but a mobile device on a commute
Date and time of user visit(s)	• When does the customer have free time to browse websites? • Does the customer go to school or have a job? The company can time advertisements or promotional emails for when they think the user will be most receptive.
Which pages were visited	• Which products interested the customer • Combined with time data, a longer page visit could indicate greater interest (or it could indicate a problem)
Words typed in a search box	• Which products interested the customer • Whether the search term matches what the product is called on the website • To help the company choose keywords to advertise products
IP address	• To find out where the customer lives, so that relevant information can be provided, e.g. shipping rates to their country

Task B. Interrogate the data

1. Which page was the first one visited on 8 May 2023? **Animal Trading Cards**
2. How many times did users search on the website? **34**
3. How long was the longest visit? **320 seconds (5:20)**
4. Which page was the most popular in terms of visits? **Flying Dragon Kite**
5. What was the mean average number of seconds per visit? **72 seconds**
6. How many different people added an item to their cart? **2 (one person added two items)**
7. People from which country visited the site the most? **Australia**
8. Draw a bar chart of the number of searches done by people from each country.

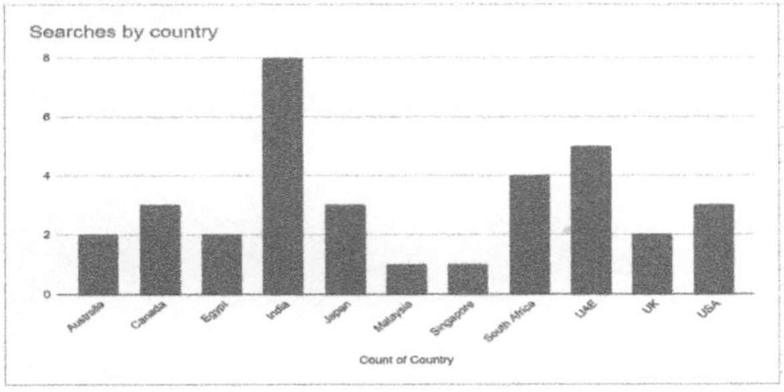

Homework

Why is it difficult to determine which item was the most searched for?

- Not all things people searched for were exact names of products
- Sometimes people make spelling mistakes in their search
- Sometimes people use a different name when they are searching for a product (e.g. someone searched for 'pirates' and someone searched for 'treasure map')

Can you come up with an idea for how you could measure which item was the most searched for, using the data available?

- You could look through the data and give each search term a label according to what you think it was searching for. Then you could count up searches with the same label to see which was the most popular.
- You could look at all of the individual words inside a search, decide which are keywords (i.e. discard words like 'and') and count up how many times each keyword is mentioned

Chapter 1.4 Models

Task A. New lunch product

There are many correct answers, here is an example:

	Option 1
Price	1.90
Number sold	10
Overall cost	14.40
Revenue	19.00
Profit	4.60

Option 1

Ingredients	Pack cost	Servings in pack	Portion cost
Rice	1.90	10	0.19
Peas	1.60	8	0.20
Tofu	2.20	4	0.55
Hoisin sauce	2.00	4	0.50
		TOTAL COST	1.44

Chapter 1.5 Create an advert

Homework

1. Explain two ways in which this technology could benefit people

- Can generate images extremely quickly, for example to prototype a product
- Free or extremely cheap, compared to a human creating the image
- Can allow people without artistic skills to create beautiful images

2. Explain two ways in which this technology might cause problems

- It can generate very realistic images of things which don't exist, so it could be used to trick or scam people
- It could cause people to stop learning art skills if computers are able to generate any kind of image on demand
- It could cause graphical artists or designers to lose their jobs
- The AI image generator could generate an image which contains copyrighted content

Chapter 1.6 'What if' analysis

Homework

1. =((B6/C6)*D6)+E6

2. Banana loaf and cheese biscuits, because their packaging price is doubled.

3. Change cell C2 to the value 6

4. The cost (2.47) would now be divided by 6 instead, meaning that the new pack cost price for brownies would be 0.86. (The brownies have still been made so the cost remains the same).

5. The calculation will give the correct result, but the baker has introduced an absolute value (0.08). This means that the value of the packaging price in the cell is ignored in the calculation.

She should have updated the packaging price and referenced cell E4 instead, the correct formula should be:

=((B4/C4)*D4)+E4

(Students have studied cell referencing in Stage 6).

Chapter 2 – Content creation

Chapter 2.1 Know your audience

Task A. Features of a virtual tour

Images	• To show what the location looks like • To provide a close up of a particular item, e.g. an interesting artefact in a museum virtual tour
Text	• To provide more detailed information about the location or about a particular item
Video	• To be able to present the location whilst also providing an audio commentary about what the viewer is seeing • To be able to show something which moves
Sound	• To provide a spoken commentary • To allow the visitor to experience sounds at the location, e.g. birds singing • So that people who are blind can also access the tour
Buttons	• To allow the visitor to move through the virtual tour • To allow a choice of where to visit next
Maps	• To show an overview of what the tour covers • To allow the visitor to click on a particular place they would like to visit

Task B. When is it appropriate?

Situation	Emoji	Meme	GIF
An instant message to a friend	✓	✓	✓
An instant message to your grandparent			
A social media post	✓	✓	✓
An invitation to a party	✓		
A History assignment for school			
An article for the school newspaper	✓	✓	

Chapter 2.2 Document templates

Task A. What are the benefits of using a document template?

The template style is automatically applied to the whole document	If you need to change something, you only need to change it once in the template
You can use a template that someone else has designed	Allows anyone to create a document that looks good
Styles are the same everywhere they are used	The document looks consistent
Documents can be set up to match a company's branding	The document always looks professional

Chapter 2.3 Searching for information

Task B. Web quest

What sweet treat is the city of Bath, UK famous for? - **Bath bun**

For other questions, the answers will depend on your local area, so be sure to search for the answers in advance of the lesson.

Chapter 2.4 Using multimedia

Task A. What is it used for?

An interview with the author of a children's book
Sound or video would be appealing if the interview is intended to be watched by young children who may not be able to read. However, text and images could be used if the interview is to be printed in a magazine.

A recipe
Text and an image would be the most common way of presenting a recipe, so that a cook can see what the finished dish looks like and easily read the ingredients whilst they are cooking. Some TV shows have video footage of famous chefs cooking, but it is more difficult to note down the ingredients.

A catalogue for the products a business sells
Text and images would be good for a catalogue that is accessed directly by customers. This might be either on paper or as part of a website. However, items which are more complicated might benefit from having an accompanying video to show what they can do.

A tutorial explaining how to make something
Online video tutorials are very popular with younger people. They can be sped up or slowed down and re-watched over again if something is particularly difficult. Older people are more used to tutorials involving text and images, but these can be less helpful than a video if the item being made is very complicated.

1. You are recording a video for a virtual tour of your school. During the recording, a teacher of another subject walks past with their class.
   - The teacher, or one of the other students may object to being filmed
   - There may be too much background noise
   - You may cause disruption by blocking a route through school
   - The teacher may ask why you are not in your classroom
   - **Stay safe** - Check that the area you are recording in is empty and not likely to be used by other people, use signs to indicate that you are filming. If you accidentally film someone without permission, ask for their permission or if this is not possible, immediately delete the footage.
2. You are taking photographs to use in your tour. Your partner takes a photograph of you in a room that is part of the tour, on their mobile phone.
   - Taking photos on personal devices may be against the school rules
   - You may not like the image that was taken, but you do not have control of it
   - Your partner may use the image for something other than the work
   - **Stay safe** - Know what your school rules are for mobile devices, and follow the guidelines agreed by the class. Understand that once an image has been taken on a device that you do not own, you no longer have control of who sees it.
3. You are recording a video of yourself visiting a local landmark, as part of your virtual tour. Your partner wants to see the video and suggests you upload it onto a video streaming site so that they can see it.
   - Anyone on the internet can see what you look like, and which area you live in
   - If you are wearing uniform, they can also see which school you go to
   - **Stay safe** - Do not upload videos which could personally identify you. Any videos shared publicly on the internet could be viewed by anyone. Share the video with your partner safely during the lesson using the school network, or allow them to watch it on your device.

Chapter 2.5 Presentation software

Homework

	Advice
Size of buttons	• Buttons on the tour should be sufficiently large for people with motor impairments to be able to use them • Making them larger will also benefit people using touch screens in general
Text contrast	• The text colour should have a high contrast with the background colour • There are free tools available to check the contrast level between a pair of colours is accessible to partially sighted people
Colour choice	• Some users may be colour blind (have colour vision deficiency) and struggle to distinguish between certain colours, e.g. red and green • You should never use colour on its own to indicate meaning - e.g. use a colour AND an icon
Font choice	• Some people with dyslexia may struggle to read certain fonts. There are special dyslexia-friendly fonts, but choosing any simple and clear font is also helpful.

Chapter 2.6 Augmented reality

Task A. What are the benefits and drawbacks of augmented reality?

	Benefits	Drawbacks
Education	You can explore and learn about things that would be too difficult/ expensive/ dangerous to explore in real life. It can help understanding of difficult concepts through interactive exploration.	It requires expensive devices to be able to use it, which might worsen the digital divide. Making AR apps is expensive and they only cover a small part of the curriculum.
Entertainment	Augmented reality apps which involve going to real-world locations might promote good health because people get exercise walking to the locations. You could make friends with other people playing the game, or play it with a friend.	Accidents may happen if people are looking at the augmented reality content and not the real world. There are privacy issues – should an app be able to know exactly where you are and what you can see?

Chapter 3 – Create with code 1

Chapter 3.1 Familiarisation with Python

Task B. Recreate the Scratch program

```
print("Hello")
answer = input("What is your name?")
print("What a cool name")
```

Homework

Code	Answer
print(name + " likes " + subject)	Maryam likes Science
print(subject + " is " + name + "‘s favourite subject")	Science is Maryam’s favourite subject
print("I like blue. " + name + " likes " + colour)	I like blue. Maryam likes blue
print("Favourites: " + subject + ", " + colour)	Favourites: Science, blue
print("Maryam’s clothes are" + colour)	Maryam’s clothes are blue (No space was added after the word ‘are’ inside the quotation marks, so the value ‘blue’ will follow immediately after. This is a common mistake students will make!)

Chapter 3.2 Program library

Task A. Video game program library

Students may come up with different answers, here are some suggestions:

- Make a character jump
- Collect an item
- Display a menu screen
- Physics for gravity acting on an object
- Move up or down a ladder
- Push an object
- Move the camera in 3D

Task B. Guess what this code does

Run the program a few more times, then explain in your own words what this program does

Answer Generates a random integer (whole number) between 1 and 6

Homework

Code	Value	Possible?
import random result = random.randint(1, 10)	1	✓
	6	✓
	0	✗
	1.5	✗
import random result = random.choice(["apple", "pear", "orange", "banana"])	"pear"	✓
	"kiwi"	✗
	"choice"	✗

Chapter 3.3 Decomposition

Task B. What's in a name?

Fearless	Nazir	Musician	of Lavaland
An adjective	A given name	A job	A place

Task A. Constants

A constant is defined in one place, usually at the **start** of the program. If you need to change the value of the constant in the code, you need to find and replace **one** value to update the whole program.

Constants are clearly distinguished from variables by putting their names in **uppercase**. The **value** of the constant is then used in the code instead of its **name**. Anyone reading the program has a clear indication of what the constant represents.

Here is some code that **does not** use a constant.

```
total = price * 0.8
```

You could use a constant called DISCOUNT to hold the value **price**, which would make the meaning of the code much clearer.

Homework

```
import random

# Define a constant value for each side of the coin
GAZELLE = 1
WRITING = 0

# Create a constant list containing the two possible sides
SIDES = [GAZELLE, WRITING]

# Simulate flipping the coin 5 times by choosing
# a random item from the SIDES list
coin1 = random.choice(SIDES)
coin2 = random.choice(SIDES)
coin3 = random.choice(SIDES)
coin4 = random.choice(SIDES)
coin5 = random.choice(SIDES)
# Add up the results to calculate how many were deer
total = coin1 + coin2 + coin3 + coin4 + coin5
# Convert the total to a string so you can concatenate it
total = str(total)

# Print out how many of the coin flips resulted in a deer
print("You got " + total + " gazelle")
```

Task A. Design your story

```
import random

ADJECTIVES = ["Fearless", "Cheerful", "Friendly", "Scary", "Kind", "Terrible"]
name = input("Please type in your name: ")
JOBS = ["musician", "builder", "trader", "farmer", "cleaner"]
PLACES = ["Lavaland", "Azimuth", "Coalville", "Yikazan"]
QUEST_ITEMS = ["golden necklace", "precious gem", "fluffy kitten"]

description = random.choice(ADJECTIVES)
job = random.choice(JOBS)
place = random.choice(PLACES)
item = random.choice(QUEST_ITEMS)

print("This is a story about")
print(description + " " + name + ", " + job + " of " + place + ".")

print(name + " was fed up with being a "  + job)
print( name + " decided to leave " + place + " and go on a quest to find a " + item)
```

Task B. What does the error message mean?

Here are some examples of error messages you might see in Python. Write down underneath each one what the error message means, and what you should do to solve it.

```
print( name + " decided to leave " + plac)
```

Python error: NameError: name 'plac' is not defined.
Answer: Python can't find a variable called 'plac'. This is probably because there is a variable called 'place' but you typed its name incorrectly.

```
print(name  " was fed up with being a "  + job)
```

Python error: Invalid syntax
Answer: There is a + missing between the variable 'name' and the start of the string. Add a + between name and the first quotation mark.

```
print("He had a quest to find a " + item)
```

Python error: TypeError: can only concatenate str (not "list") to str
Answer: The variable 'item' contains a list. This is likely to be because you have created the list but forgotten to choose a random item from it.

(Note that this kind of error is easier to spot if students create their lists as constants, as the name will be in capitals!)

Homework

```
import random

ADJECTIVES = ["fearless", "cheerful", "friendly")

place = random.choice(PLACES)

PLACES = ["Lavaland", "Azimuth", "Coalville"]
ENEMIES = ["mammoth", "giant spider", "large sheep"]
name = input("Please type in your name: ")

description = random.randint(ADJECTIVES)

enemy = random.choice(ENEMES)

print("There was once a " + description + " person called "  name)
print("One day, a " + enemy + " came to " + place)
```

Answers (in order of highlighted mistakes)

1. Wrong type of bracket – change it to a square bracket.
2. You can't choose a place from a list that doesn't yet exist. Move the whole line of code after the list of places.
3. Wrong function from the random library: randint selects a random number. Change the code to use the choice function.
4. ENEMIES has been spelled incorrectly, change the spelling.
5. There is a missing + sign between the quotation mark and the name variable.

Chapter 3.6 Feedback on software

Task B. Review the code

Less experienced developers can improve their skills after receiving feedback from a more experienced developer.

- The overall quality of the code is improved
- Another developer may come up with an improvement (e.g. that makes the code run faster or more efficiently)

Task C. User feedback focus

- The user may not use the software in the way the developer thought they would, meaning the developer may need to make some changes
- The developer can see everything the user does, so they do not have to rely on only what the user chooses to (or remembers to) tell them about
- Speaking to the developer may cause the user to forget what they were doing, or to use the software in an unnatural way because they feel like they are being tested

Chapter 4 – How computers work

Chapter 4.2 The operating system

Task A. Which tasks are performed by the operating system?

Plays music *application software*	**Not OS**	Protects the computer against viruses	**OS**
Allocates the system's memory	**OS**	Allocates processing time	**OS**
Displays your files so that you can find and access them	**OS**	Lets you play a video game *application software*	**Not OS**
Records a voice note *application software*	**Not OS**	Provides software to create a presentation *application software*	**Not OS**
Stores the instructions required to start up the computer *ROM*	**Not OS**	Communicates with hardware devices	**OS**

Chapter 4.3 Utility programs

Task A. Purpose of utility programs

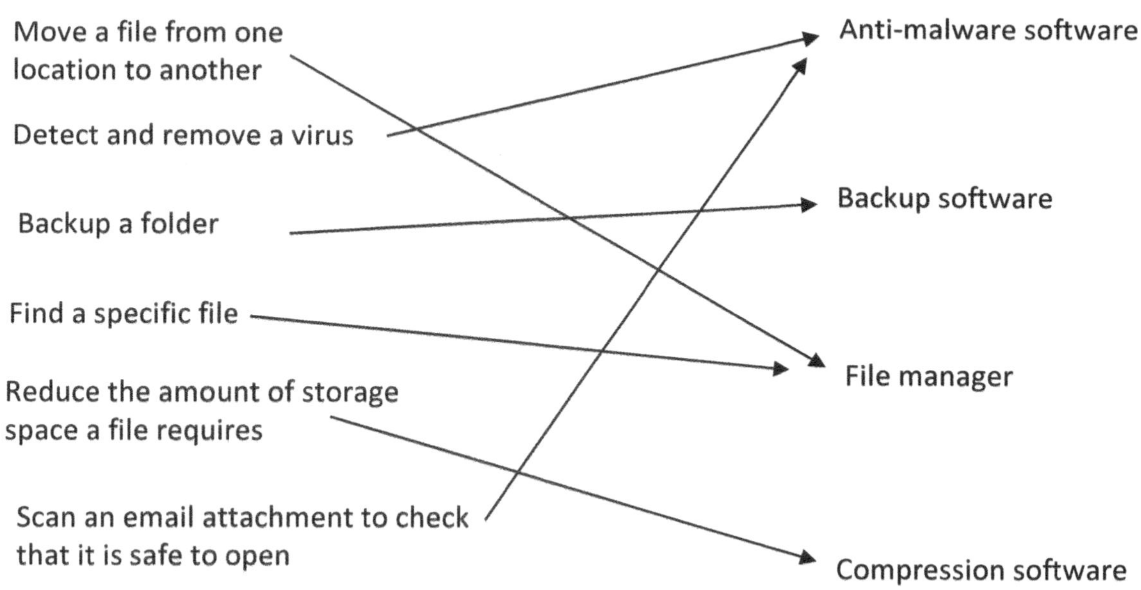

Task B. Advice from a technician

1. You could scan the file using antivirus software to check whether it is safe to open. If you are able to, please contact your boss via another method, for example by telephone, to check whether the email is genuine.
2. It sounds like you have been a victim of a ransomware attack. Please bring your computer to the IT support department and I may be able to restore your files from a backup. It is important that this computer is quarantined from the network to avoid spreading the ransomware to other people on the network.
3. The network has antivirus and antispyware software running, which will work to detect any viruses attempting to enter the network. It is automatically updated via the technical support department, rather than having to be updated by each individual employee.

Chapter 4.4 Unauthorised access

Task A. Simulate a firewall

Fill in the table as you do the role play exercise.

Task	Result
Access the website at 192.168.3.76 port 443 from inside the trusted network	Direction: Out (for the request) Destination: 192.168.3.76 Port: 443 Decision: Allow Direction: In (for the data) From: 192.168.3.76 Port: 443 Decision: Allow **ALLOWED**
Access the website at 43.98.2.154 port 80 from inside the trusted network	Direction: Out (for the request) Destination: 43.98.2.154 Port: 80 Decision: Block **BLOCKED** by the firewall, before the request reached the web server.
Open a connection from the internet to port 22 on the trusted network	Direction: In Destination: 22.36.4.11 Port: 22 Decision: Block **BLOCKED** – connection rejected
Open a connection from inside the trusted network to 64.63.2.21 port 22 on the internet	Direction: Out Destination: 64.63.2.21 Port: 22 Decision: Block **BLOCKED** – port not allowed
Open a connection from inside the trusted network to 64.63.2.21 port 300 on the internet	Direction: Out Destination: 64.63.2.21 Port: 300 Decision: Allow Direction: In Destination: 64.63.2.21 Port: 300 Decision: Allow **ALLOWED**

Task B. Keeping data safe inside a network

1. It depends. Some firewalls can prevent email attachments being sent – they are often set up to reject a certain file type, or to block attachments sent to an external source. However, it would depend on how the firewall was set up, as emailing documents outside the organisation is also a legitimate thing that an employee might need to do.
2. Faisal probably should not have had access to this data. It says in the activity that he is a software engineer, so details of all clients would not be required to do his job. If he had been, for example, in the sales or accounts teams, he may have needed access to the data.
3. File access permissions should always be set up as the minimum access required for a person to do their job. Since Faisal did not need this data for his job, he should not have had access to it.

Chapter 4.5 Network connections

Task A. Advantages and disadvantages of wired and wireless networks

	Wired	Wireless
Connection speed	+ Faster connection speed than wireless	- Connection is slower than wired - Might be subject to interference (e.g. in a thunderstorm) or poor signal strength (e.g. through thick walls)
Security	+ Need to be physically inside the building with a cable to use it	+ Modern encryption protocols are very good - If not properly secured, the network can be accessed by anyone, even outside the building
Other	- Cables can get damaged or tangled, or look unsightly - More expensive to set up due to cost of wiring - Need a specific port on your device to use it, i.e. can't be used with a smartphone	+ Can move around freely without having to worry about cables + Can be offered in a cafe to the public

Task A. Test the troubleshooter

Scenario	Advice given	Evaluation
1	The person is using a wireless connection, which is running very slowly. Since programs are opening OK it is unlikely to be a problem with RAM. The advice could be to move closer to the wireless access point, or to check that the wireless router is set up correctly with a strong password as someone may be using the connection without permission.	
2	There are two options here. One option is to upgrade the amount of secondary storage space available, e.g. by adding either an additional hard drive, or by saving some of the files onto an external device such as a USB flash drive. Another option would be to use a utility program to compress the files and thus reduce the amount of storage space they take up.	
3	It is likely that the person does not have permission to modify the file. Perhaps it is stored in a folder or drive that can only be modified by the other department. Investigate the file permissions for that file, and if necessary, give them permission to modify the file.	
4	It is likely that the computer is struggling with the number of pieces of software that are open at once, given the RAM it has available. Advise the user to close some of the applications so that they do not have too many pieces of software running at once. If this is a task the user might have to do regularly, it might be good to look into whether you can upgrade their RAM to a larger capacity.	
5	It is likely that there is something wrong with the wired connection, for example an ethernet cable may have been unplugged. Check that the ethernet cable is plugged in, and that no cables have been damaged.	

Chapter 5 – Create with code 2

Chapter 5.1: Data types

Task A. What type of data?

1. String
2. String
3. Integer
4. Boolean
5. String, so that it can be concatenated with other strings to be printed
6. Line 8 has a mistake. The code str(points) * 3 converts the data in the points variable to a string, and then multiplies it by 3. So if the points variable contained the value 5, this would become "5", and "5" * 3 = "555".

   To fix this, the result of the *whole* calculation should be cast to a string instead:

   str(points * 3)
7. print(Kofi + " scored " + str(points * 4) + " points!")

Homework Recap quiz

1. Integer: any whole number e.g. 1, 2 etc.
   String: one or more characters in quotation marks, e.g. "hello"
   Boolean: True / False (no quotation marks, capital first letter)
   Real: any number with a fractional part e.g. 3.1415
2. amount = int(amount)
3. Casting means changing the data type of a variable, for example from a string to an integer.
4.
   - If you are asking the user to input some data which is a number/integer/real number
   - If you need to concatenate a variable of a different type into a string to produce an output.

Chapter 5.2 Pseudocode

Task A. Similarities and differences

Similarities

- Both use similar language e.g. print, input
- Both are followed in order from top to bottom (this is not true for algorithms involving selection/iteration)
- Both have clear unambiguous instructions

Differences

- Flowchart symbols provide meaning, e.g. input/output or a process
- Flowchart has a clear start and end marked
- Pseudocode takes up much less space
- Pseudocode looks much more like an actual program

Task B Flowchart to pseudocode

```
PRINT "Hello, what is your name?"
name = INPUT
PRINT "Hi " + name
```

Task C. Pseudocode guidelines

You may wish to add guidelines of your own, but here are some you could suggest:

- Use capital letters for keywords e.g. PRINT, INPUT.
- Always use "quotation marks" around string data.
- Make sure each statement is short, clear and precise.
- Make clear where the pseudocode starts and ends.
- Think about what code you might use when you write a statement.
- Write clearly and legibly.
- You don't need to worry about casting in pseudocode.

You may wish to add this later on in this chapter as it will become relevant when students need to write pseudocode with conditional/iteration statements

- Use indentation to show the logic of the program.

This is not the only correct solution – it is an example.

```
PRINT "Hello, what would you like to buy?"
item = INPUT string
PRINT "Please enter the price of the scarf"
price = INPUT float
PRINT "What is the % discount?"
discount = INPUT integer
result = price - (price * (discount/100))
PRINT "A" + item + "with a" + discount + "% discount costs " + result
```

Homework

Type in the numbers Add up the first and second number Add up the third and fourth number Multiply them together Print the result	• The statements are not precise enough – how many numbers do they need to type in? Which numbers need to be multiplied together?
PRINT "Please type in a fruit" INPUT fruit PRINT "Please type in another fruit" INPUT fruit	• Each of the fruits needs to be stored as a separate variable otherwise in a real program the value of the first fruit would be overwritten by the second
PRINT "Enter a number" result = calculate the number squared PRINT "Your number squared is " + result	• It is not clear from the pseudocode how to calculate the number squared. You may know how to do this, but will someone else?

Chapter 5.3 IF, ELIF and ELSE

Task B. Evaluate the conditions

1. true
2. false
3. true
4. true
5. true
6. false

Homework

1. Input = 40 – **Well done**
2. Input = 51 – **Pretty good**
3. Input = 99 – **Pretty good**

For number 3, the input of 99 matches the first condition as 99 > 50, so the flow of control executes the line `print("Pretty good!")` and then skips the rest of the statement.

Chapter 5.4 AND, OR, NOT

Task A. Recap logic gates

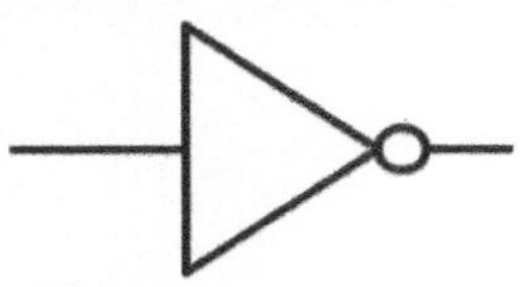 NOT gate	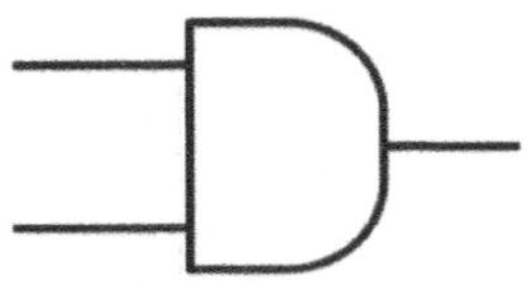 AND gate	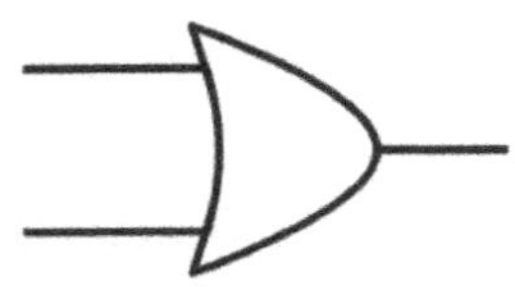 OR gate

Task B. Complete the truth tables

1.

A	B	Output
1	1	**1**
1	0	0
0	1	0
0	0	**0**

2. Only when both of the inputs are true (1)

3.

A	B	Output
1	1	**1**
1	0	**1**
0	1	**1**
0	0	**0**

4. When either one or both of the inputs are true (1)

Homework

1. Write down all of the possible combinations of inputs A and B in the table.

A	B	C	Output
1	1		
1	0		
0	1		
0	0		

2. For each row, look at A and B, and work out what the output at C would be for that combination. Write your answers in the table.

A	B	C	Output
1	1	1	
1	0	0	
0	1	0	
0	0	0	

3. Finally, using the value you worked out for C as the input to the NOT gate, write down the output for each row.

A	B	C	Output
1	1	1	0
1	0	0	1
0	1	0	1
0	0	0	1

4. The output is true when neither input is true.

Discuss

A barrier to exit a car park that will lift when a car is present and a ticket is inserted:

AND gate

Car?	Ticket?	Output (barrier up)
1	1	1
1	0	0
0	1	0
0	0	0

A restaurant that does not allow people in if they are wearing trainers: NOT gate

Trainers?	Output (allowed in)
1	0
0	1

A vending machine that dispenses a drink if money has been inserted and a button has been pressed: AND gate

Money?	Button?	Output (drink dispensed)
1	1	1
1	0	0
0	1	0
0	0	0

A fridge that sounds an alarm if the door is left open or if the temperature goes above 8°C: OR gate

Door open?	High temp?	Output (alarm?)
1	1	1
1	0	1
0	1	1
0	0	0

Build

```
username = input("Enter your username: ")
password = input("Enter your password: ")

if username == "zoya" and password == "apple32":
    print("Welcome " + username)

else:
    print("Sorry, you are not allowed in")
```

Build

```
username = input("Enter your username: ")
password = input("Enter your password: ")

# Tip: For long conditions, put the whole condition in parentheses
# so that it can be separated onto multiple lines
if ((username == "zoya" and password == "apple32")
or (username == "admin" and password == "secret")):
    print("Welcome " + username)
else:
    print("Sorry, you are not allowed in")
```

Chapter 5.5 Design and build a quiz program

Task A. Finish the pseudocode

Here is a possible answer, although the pseudocode can be in any format as long as the meaning is clear.

```
INPUT q1answer1
INPUT q1answer2

IF (q1answer1 == "A" AND q1answer2 == "B") OR
  (q1answer1 == "B" and q1answer2 == "A")
   OUTPUT "Correct!"
ENDIF
```

Homework

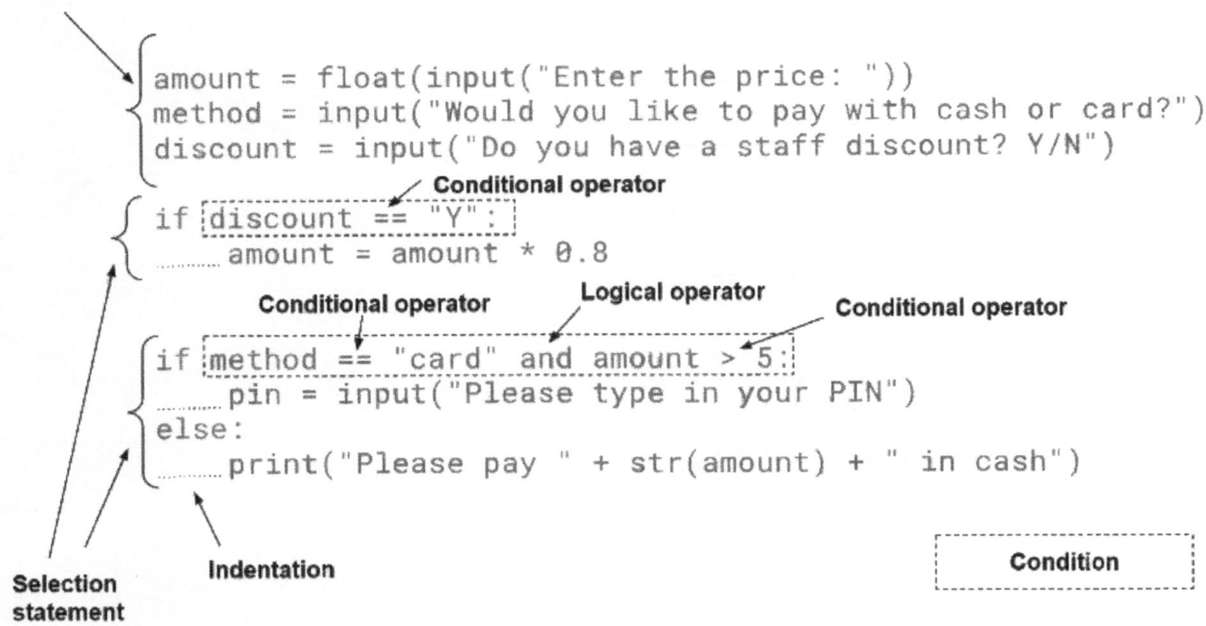

Chapter 5.6: Showcase your quiz

Here is an example quiz with three questions, to demonstrate the use of all three logical operators.

```
print("Welcome to the quiz!")
score = 0

# Add your questions here
# ----------------------------

print("Which two animals have wings?")
print("A - Bat")
print("B - Parrot")
print("C - Rhinoceros")
print("D - Tiger")
q1answer1 = input("Enter A, B, C or D for the first animal: ")
q1answer2 = input("Enter A, B, C or D for the second animal: ")

if (q1answer1 == "A" and q1answer2 == "B") or (q1answer2 == "A" and q1answer1 == "B"):
   print("That was correct! Well done")
   score = score + 1
else:
   print("Oh no, you got it wrong :(")
```

```
print("Give one continent where bears are a native species")
print("A - Europe")
print("B - Africa")
print("C - Asia")
print("D - Australia")
q2answer = input("Enter A, B, C or D: ")

if q2answer == "A" or q2answer == "C":
   print("You know a lot about animals!")
   score = score + 1
else:
   print("Bad luck, that was the wrong answer")

print("Which of the following animals is NOT a herbivore?")
print("A - Snail")
print("B - Elephant")
print("C - Horse")
print("D - Rabbit")
q3answer = input("Enter A, B, C or D: ")

if not q3answer == "A":
   print("Whoops, that wasn't right")
else:
   print("Incredible animal knowledge!")
   score = score + 1

# End of quiz
# ----------------------------
print("Congratulations! Your score was " + str(score) )
```

Homework

1. A AND B
2. A OR B
3. NOT A
4. NOT B

Chapter 6 – Connect the world

Chapter 6.1 Binary and denary

Task A. Binary to denary conversion

1. 11111 = 31
2. 00111 = 7
3. 01100 = 12
4. 10001 = 17
5. 11010 = 26

Task B. Denary to binary conversion

1. 2 = 00010
2. 18 = 10010
3. 23 = 10111
4. 30 = 11110
5. 4 = 00100

Homework

1. 8410 = 10101002
2. 1010 = 10102
3. 102 = 210
4. 111101002 = 24410
5. 25510 = 111111112
6. 000010112 = 1110
7. 10010 = 11001002
8. 10002 = 810
9. 3910 = 1001112
10. 1001012 = 3710

Chapter 6.2 How are we connected?

Task A. Copper and fibre optic cables

1.

Type of cable: **Copper cable**

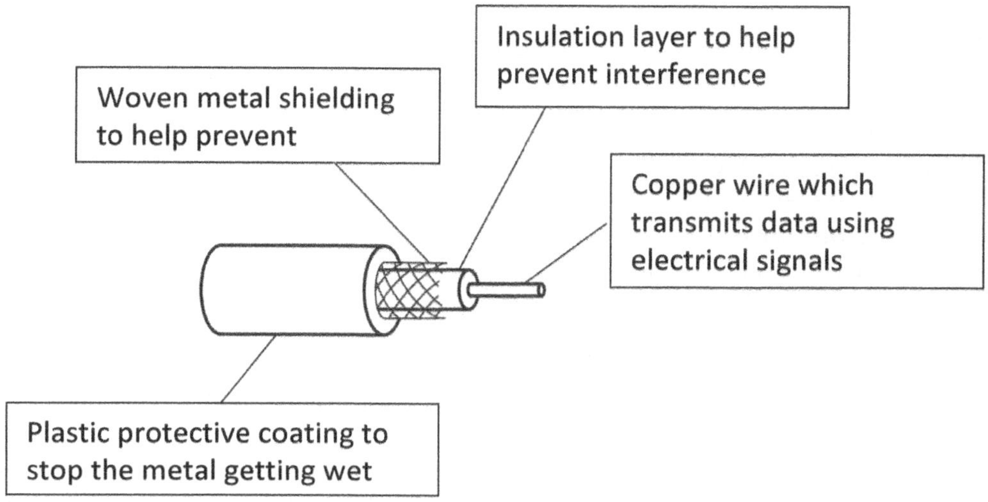

Type of cable: **Fibre optic cable**

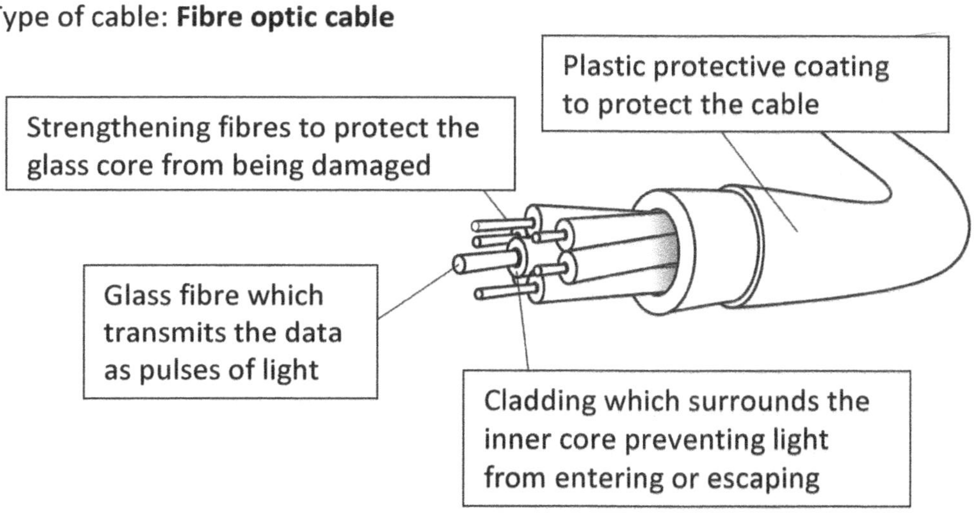

2.

	Advantages	**Disadvantages**
Copper cable	• Copper is cheap and easy to obtain	• Signal can be disrupted by interference • Cannot transmit data as far as fibre optic cable before the signal needs to be boosted
Fibre optic cable	• Can transmit over a very long distance • Data transmits very quickly • Strong and flexible Immune to electromagnetic interference	• More expensive to install than copper cable (but requires less maintenance) • Joining cables together requires special equipment

Task B. PAN, LAN or WAN?

Statement	PAN	LAN	WAN
Connects devices over a large geographical area			✓
Connects devices in one geographical location	✓*	✓	
Formed by connecting together multiple local area networks			✓
Allows you to communicate across continents			✓
An example is a network owned by a business		✓	
An example is the internet			✓
Connects multiple different devices	✓	✓	✓
Used to connect a wireless mouse to your laptop	✓		

*A PAN and a LAN both connect devices in one geographical location, but the scope of a PAN is limited to those devices in range of a single user whereas a LAN covers an entire building.

Chapter 6.3 ASCII

Task A. Write your name using ASCII

Letter	Denary	Binary
L	76	0100 1100
a	97	0110 0001
u	117	0111 0101
r	114	0111 0010
a	97	0110 0001

Homework

ASCII code	Denary	Letter	ASCII code	Denary	Letter
0101 0111	87	W	0110 0100	100	d
0110 0101	101	e	0110 1111	111	o
0110 1100	108	l	0110 1110	110	n
0110 1100	108	l	0110 0101	101	e
0010 0000	32	Space	0010 1110	46	.

Chapter 6.4 Online communication

Task A. Choose and justify a communication method

1. If you have the customers' email addresses and permission to contact them in this way, you could send them a promotional **email** to alert them to the special offer. This would be a good choice because customers are used to being contacted via email for promotions, and they can read the email at a time that is convenient to them.

   Alternatively, you could use a **social media** post to advertise your promotion. This would be a good choice because it can potentially reach a large amount of people, and the post can be shared so that friends of your customers will also see the details of the offer.

2. Probably the best solution in this situation would be a **video call** so that you can see the patient face to face and explain how to do the exercises in person. However, this may not be possible if the elderly person does not have access to a device with a camera.

   Alternatively, you could send an **email** to the person with a document attached explaining how to perform the exercise. This would be a good choice because many older people know how to use email, or you may be able to email the document to a relative who could help, if you have permission.

3. If the **online game** you are playing has voice chat capabilities built in, this would be a good choice because it would mean that all players would have access. Using the voice chat option would be better than using text chat because it is much quicker to speak than type a message whilst playing a game.

   Alternatively, if there are no built-in capabilities, you could use a separate **voice chat** server to communicate with the other players. This would allow you to coordinate verbally whilst playing the game, and moderation features would also mean you could control who can or cannot participate in the chat.

Homework

	Online	Offline
Sharing complicated information	• Can explain complicated things live via video stream • Possible to provide images and diagrams • Can send large text documents as attachments	• Can give a demonstration if you are in the same place • Could print a document and send it via mail, but this would take longer to arrive
Miscommunication	• More likely for text-based communication as there is no tone or facial expression • Can help to avoid this by using emojis	• Less likely when talking to someone in person as they can see your expressions
Cost	• Generally free, assuming you have an internet connection • Some services have subscription fees • A device is required which may be expensive	• Free to communicate • Transport costs to be in the same location • May cost more in terms of time - e.g. have to travel to a meeting • Postage fees

Chapter 6.5 Data transmission

Message 1
W1B2W1B2W1B1R2B1R2B1R14B1R5B1W1B1R3B1W3B1R1B1W2

Message 2
W2G1W2G1W2G1B1G2B1G1W2G4W2G2R2G1W1G6W1G6W1G1W1G1W2G1W1

Message 3
W1B1W3B1W1B1Y1B1W1B1Y1B1W1B1W2G1B1W2G1W1G2W3G1W1G1W5G2W6G1W3

Message 4
W1Y4W2Y1B1Y2B1Y1W1Y6W1Y1R1Y2R1Y1W1Y2R2Y2W2Y4W9

Message 5
W4R2W4R1W4B1W1R1B1W3B1W1R1B1W3B1L2B1W3B1L2B1W3B4W2

Message 6
W3P1W5P2R1W3P1R1P3W3O3W4O3W4O3W5O1W3

Message 1 – heart

W	B	B	W	B	B	W
B	R	R	B	R	R	B
R	R	R	R	R	R	R
R	R	R	R	R	R	R
B	R	R	R	R	R	B
W	B	R	R	R	B	W
W	W	B	R	B	W	W

Message 2 – frog

W	W	G	W	W	G	W
W	G	B	G	G	B	G
W	W	G	G	G	G	W
W	G	G	R	R	G	W
G	G	G	G	G	G	W
G	G	G	G	G	G	W
G	W	G	W	W	G	W

Message 3 – flowers

W	B	W	W	W	B	W
B	Y	B	W	B	Y	B
W	B	W	W	G	B	W
W	G	W	G	G	W	W
W	G	W	G	W	W	W
W	W	G	G	W	W	W
W	W	W	G	W	W	W

Message 4 - smiling face

W	Y	Y	Y	Y	W	W
Y	B	Y	Y	B	Y	W
Y	Y	Y	Y	Y	Y	W
Y	R	Y	Y	R	Y	W
Y	Y	R	R	Y	Y	W
W	Y	Y	Y	Y	W	W
W	W	W	W	W	W	W

Message 5 – glass of water with straw

W	W	W	W	R	R	W
W	W	W	R	W	W	W
W	B	W	R	B	W	W
W	B	W	R	B	W	W
W	B	L	L	B	W	W
W	B	L	L	B	W	W
W	B	B	B	B	W	W

Message 5 – ice cream

W	W	W	P	W	W	W
W	W	P	P	R	W	W
W	P	R	P	P	P	W
W	W	O	O	O	W	W
W	W	O	O	O	W	W
W	W	O	O	O	W	W
W	W	W	O	W	W	W

Chapter 7 – The power of data

Chapter 7.1 Keeping records

Task A. Label the catalogue card

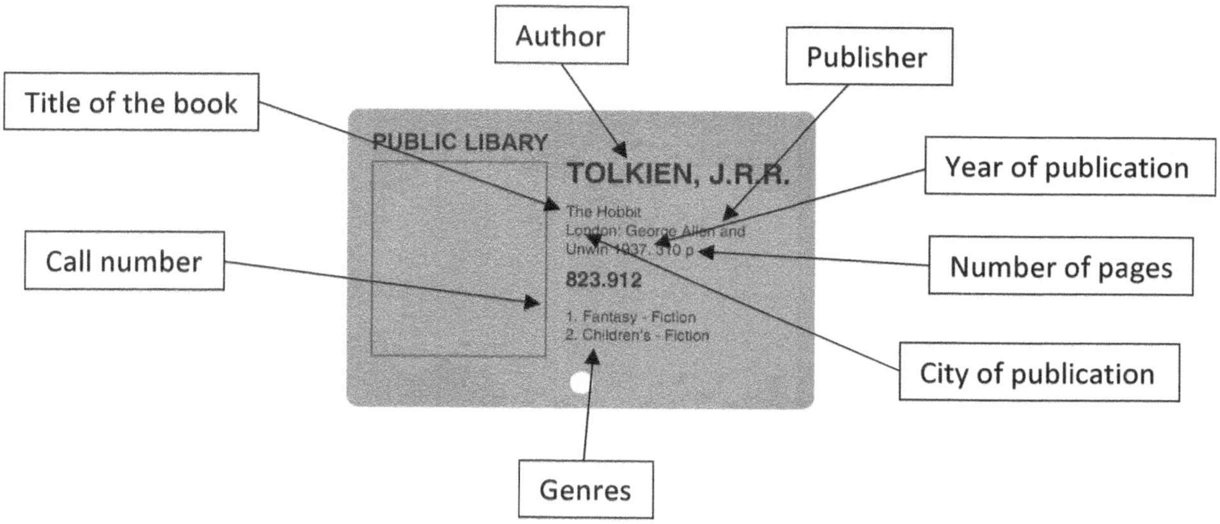

Task B. Compare library systems

Theme	Card catalogue	Modern database
What data is collected about a book?	The information labelled on the card above.	The information above, plus other information such as ISBN number, cover picture, edition, size of book, paperback or hardback
How is the data about a book stored?	Written or typed on cataloguing cards by a person.	Added to a database saved electronically.
How would you find a specific book?	Look in the correct drawer to find the card relating to a specific book. Cards were indexed by author, book title or genre.	Search on a computer by typing in any information relating to the book.
How would you borrow a book?	Take your book to the librarian who would remove the ticket inside the book and place it in the space in a drawer allocated to your library card. They would stamp the return date in the front of the book.	Scan your library card using the barcode scanner. Then, select the option to borrow a book and place the books on a scanning bed where their RFID tags will be scanned and the data will be added to a database. A receipt will be printed indicating the return date.

Homework

Card catalogue	Modern library database
• Cataloguing cards • Drawers to hold the cards • Typewriter • Pen • Returns boxes • Library card • Ticket inside each book	• Computer • Monitor • Mouse • Keyboard • Barcode scanner • RFID tags (stuck in books) • RFID scanner • Touchscreen • Returns boxes • Library card

Chapter 7.2 Linear search

Task A. Describe a linear search in words

- You will need a list of items, and an item you are searching for
- Look at the first item in the list
- Compare the item to the item being searched for
- If the item is the same as the search item, mark it as found
- If not, move on to look at the next item
- If there are no more items in the list, the search item is not in the list

Task B. Answer the questions about the trace table

1. The first item in the list is 17, and the FOR EACH line of code will look at each item in turn, starting with the first.
2. At this point, the value of search_item is 2, and the value of item is 17. 2 is not equal to 17 so the condition is false.
3. Line 7 in the code specifies NEXT ITEM which is the ending statement for the FOR EACH block, so the flow of control returns back to line 3 to look at the next item in the list.
4. The program will continue to look at all of the remaining items in the list one by one. The value of item will change from 2 to 23, then 8, then 1, but there will be no more output from the program.

Task C. Record the new algorithm

```
01  ITEM_LIST = [17, 2, 23, 8, 1]
02  search_item = INPUT item you are searching for
03  found = False
04  FOR EACH item IN ITEM_LIST
05      IF item EQUALS search_item
06          OUTPUT "Found it"
07          SET found = True
08      ENDIF
```

```
    NEXT item
    IF found EQUALS False
        OUTPUT search_item + "was not in the list"
    ENDIF
```

Homework

1. Pile 1
2. 6
3. 5
4. 26 (if the book she was searching for was at the bottom of the pile)

Chapter 7.3 Data capture forms

Task A. Would it be accepted

The data that would be accepted is in the shaded cells.

1.

Input	Validation	Data				
Text box	Length check, greater than 5	cheese	five	20	romance	189560

2.

Input	Validation	Data				
Radio buttons with labels 1–5	Presence check	2		1	b	6

3.

Input	Validation	Data				
Text box	Type check – number	999	three	2	1.5	“eight”

Chapter 7.4 Data validation

Homework

Here are some example answers. Error messages should be polite and clearly state what the user needs to do to rectify the mistake.

Data	Validation rule	Error message
Year of publication	Type check: number	Please enter a year as a number, e.g. 1956.
Author	Presence check	Please type in the name of the author.
Title of book	Length check: >3	The title of the book must be 3 or more characters long.

Chapter 7.5 Recommend a book

Task A. Draw a mind map

Mind map example

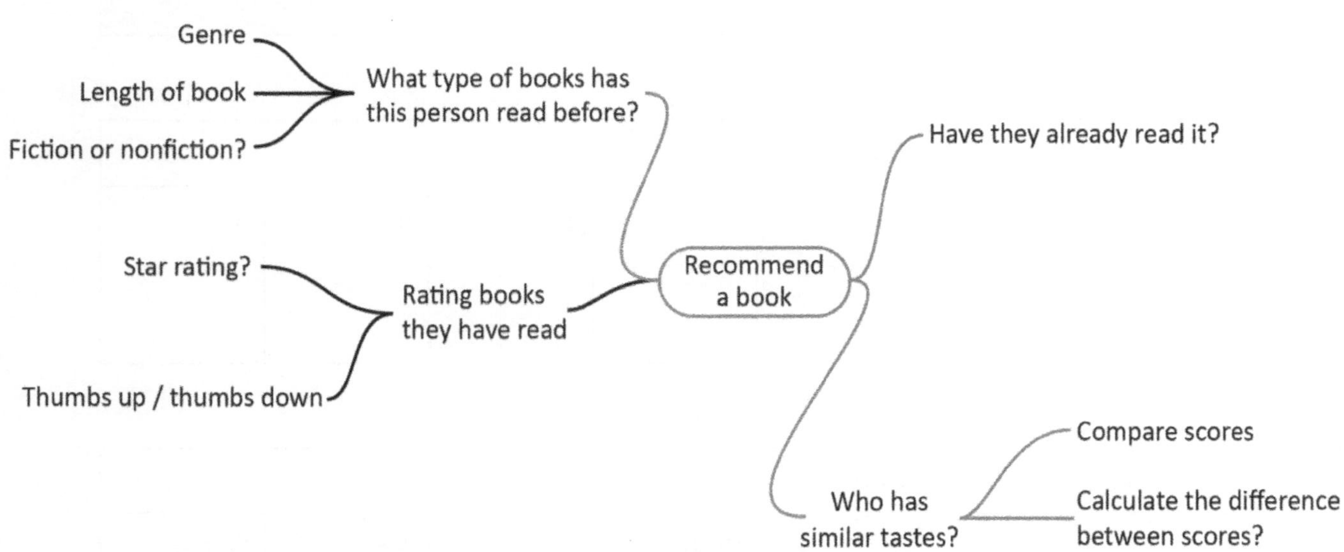

Homework

Some ideas include:

- People are asked to rate the TV show/book/product
- "People like you also liked/watched..."
- System might look at an item you bought, then look up another person who bought the same item and recommend a product they also bought
- (Relate back to metadata) Site might collect metadata about the products/shows etc. you clicked on previously and use that to remind you of those products/shows

Chapter 8 – Create with code 3

Chapter 8.2 Autonomous programming and AI

Task A. Is it autonomous?

Look at the descriptions of four different robotic systems below. Tick the correct boxes and use the final column to explain why you chose your answer.

Description	Automatic	Autonomous	Justify your answer
Parcel delivery drone	✓	✓	The drone can find its own path to the destination, including avoiding obstacles, so it is autonomous.
Drink vending machine			A vending machine requires human input to select a drink.
Machine that builds cars	✓		The machine does as it is programmed. It will not be able to change its behaviour if something unexpected happens.
A robot vacuum cleaner	✓	✓	Although it is very basic, the vacuum cleaner can decide and change its own route based on the items it finds in its surroundings.

Homework

Here is an example of an answer for the research part of the task:

AI technology I researched: Large language model (LLM) e.g. ChatGPT or Google Bard

A workplace it is used in: Job recruitment

How is it used to help people in the workplace?

Large language models can be used in recruitment to help people to showcase their skills and find a job. The recruiter might give the LLM the details about a candidate and ask it to automatically write a cover letter summarising the person's skills.

Chapter 8.3 Radio signals

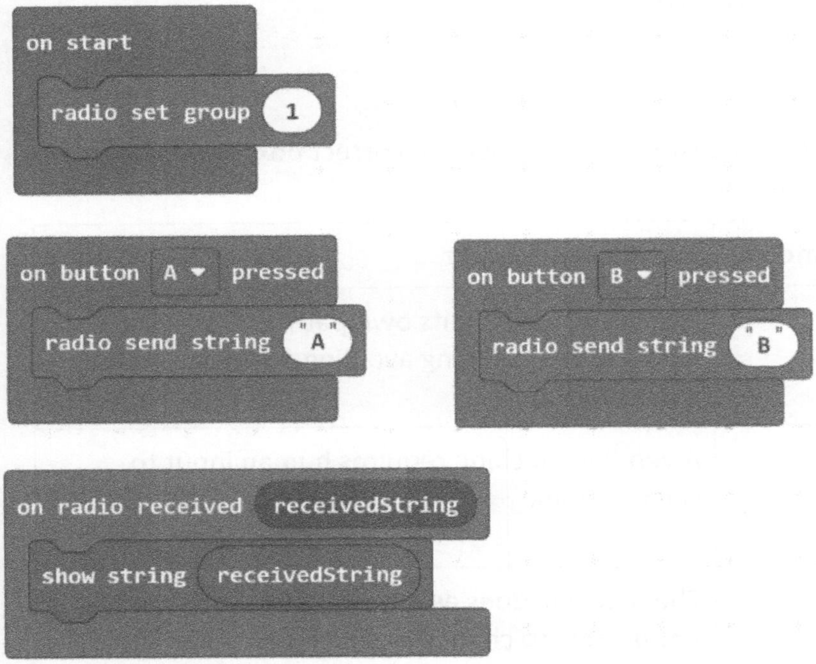

Homework

Situation	Explanation
Entering a railway platform	A card containing an RFID tag can be tapped on a reader, which allows a barrier to lift so that a person can walk through on to the platform. The tag can identify who the card belongs to so that the right person can be charged for the journey.
Entering a building	Employees are often provided with access cards containing an RFID tag, which can be tapped upon reader devices in order to enter a locked building.
Borrowing a library book	Each book contains an RFID tag with details of the book. When the book is placed on the scanner, the machine knows which books someone wants to borrow and can link these to their library account.

Chapter 8.4 Design a workplace system

Homework

Line	order_number	Input	Output	Radio
1				
2	0			
3			Square icon	
		A		
5				
6	1			
7			1	
		B		
9				
10				1
11			Tick icon	

Chapter 8.5 Test plans and test data

Discuss

It is helpful to test invalid data because it is important to test whether certain inputs which should not be allowed are properly handled by the program. Ideally, the program should either prevent the user from entering this data altogether, or it should 'fail gracefully': i.e. there should be a message telling the user that this input was not allowed, rather than the program crashing.

Homework

Test 1

- No order number has been specified as the input.
- The customer micro:bit has not been programmed with a specific order number.

Test 2

- This is a well-written test. It is clear which micro:bit to use, what the tester should do and what they are looking for in the output.

Test 3

- This test is mostly well written, the description and input boxes are clear.
- The expected output is unclear. This needs to describe what the tester should see, not what they shouldn't

Websites

Introduction

Reference	Website URL
IntR1	https://www.ijcses.org/index.php/ijcses/article/view/15
IntR2	https://doi.org/10.1186/s40594-023-00434-7
IntR3	https://stemeducationjournal.springeropen.com/articles/10.1186/s40594-023-00434-7
IntR4	https://doi.org/10.1145/3287324.3287477
IntR5	https://edublocks.org/
IntR6	https://docs.edublocks.org/
IntR7	https://edublocks.org/123-basic-set.pdf
IntR8	https://scratch.mit.edu
IntR9	https://scratch.mit.edu/faq#accounts
IntR10	https://machinelearningforkids.co.uk/
IntR11	https://machinelearningforkids.co.uk/#!/help
IntR12	https://www.google.com/docs/about
IntR13	https://support.google.com
IntR14	https://www.google.com/sheets/about
IntR15	https://www.google.co.uk/forms/about
IntR16	https://support.google.com
IntR17	https://microbit.org/get-started/getting-started/introduction
IntR18	https://microbit.org/teach/professional-development
IntR19	https://microbit.org/get-started/features/overview

Chapter 1

Reference	Website URL
C1L5R1	https://www.craiyon.com
C1L5R2	https://www.adobe.com/products/firefly.html
C1L5R3	https://www.canva.com/en_gb/
C1L5R4	https://pixlr.com

Chapter 5

Reference	Website URL
C5L4R1	https://logic.ly/demo/
C5L4R2	https://scratch.mit.edu/projects/921858880

Chapter 6

Reference	Website URL
C6L2R1	https://en.wikipedia.org/wiki/Electromagnetic_interference
C6L3R1	https://adacomputerscience.org/concepts/text_ascii_char_set

Chapter 7

Reference	Website URL
C7L1R1	https://www.usg.edu/galileo/skills/unit04/primer04_04.phtml
C7L4R1	https://isbndb.com/
C7L4R2	https://search.worldcat.org/

Chapter 8

Reference	Website URL
C8L3R1	https://makecode.microbit.org/
C8L3R2	https://makecode.microbit.org/
C8L3R3	https://microbit.org/projects/make-itcode-it/jukebox/for more details
C8L3R4	https://microbit.org/projects/make-it-code-it/proximity-beacon/
C8L4R1	https://makecode.microbit.org/

Answers

Reference	Website URL
C1L1A1	https://www.wikipedia.org/
C1L1A2	https://www.youtube.com
C1L1A3	https://www.canada.ca/en.html